On and Off
The Beaten Track

On and Off
The Beaten Track

WILLIAM TALBOT

The Pentland Press
Edinburgh – Cambridge – Durham – USA

© William Talbot, 1994

First published in 1994
by The Pentland Press Ltd
1 Hutton Close
South Church
Bishop Auckland
Durham

All rights reserved.
Unauthorised duplication
contravenes existing laws.

British Library
Cataloguing-in-Publication Data

A catalogue record for this book
is available from the British Library

ISBN 1-85821-210-3

Typeset by Carnegie Publishing Ltd, 18 Maynard St, Preston
Printed and bound by Antony Rowe Ltd, Chippenham

To Betty.

ABOUT THIS BOOK

On leaving school young William is articled as a 'Creeper' to a strict disciplinarian on a tea estate in Ceylon. Thankfully, Bill gets himself a job on an up country estate, and soon settles happily into the life. Dolly, his mother, comes out to join him.

Five years later, Bill is due for Home leave and they both return to England. Bill joins a party going to Austria for the winter sports. With other young people, Bill is captivated by the sport, the music and romance of the visit. He becomes particularly friendly with Betty, whom he invites out to a dance at the Sports Café. They see more of each other, and on the last day of his leave he proposes to her. It is three months before they become engaged, by which time the war has broken out in 1939. There follow agonizing attempts by them both to join each other, all of which are frustrated. Finally, at her third attempt to sail, Betty is shipwrecked, to be saved by another boat. Bill is finally released by the Ceylon Defence Force to be commissioned in India. Bill and Betty give up hope of marriage during the war. They continue their letters to each other which now take six weeks to arrive, though the Airgraphs are better. At times when hope fails, they desperately comfort each other.

Then a miracle happens. Bill is repatriated to England. On his way Home the exhilaration is unbearable – what will they both be like after five years' separation? What can he expect?

This book by William Talbot paints a vivid picture of young William's early life, and his first impressions of the East. His description of life on a pre-war tea estate are novel and varied. The wartime years give rise to all the human emotions of joy, love, hope, despair, determination and tenacity. It is powerful writing.

CONTENTS

	Foreword	ix
	Introduction	xi
1.	A Creeper in the Jungle	1
2.	Great Expectations	17
3.	England, Oh My England!	60
4.	You Go On To Hitler – But We Stay Here!	69
5.	Footloose and Fancy Free	91
6.	The Wheel of Fortune Turns	114
7.	Back to the Wilderness	121
8.	The Wheel of Fortune Turns Once More	159
9.	Love – Treasure Beyond Price	186
	Postscript	194

FOREWORD

BY

SIR STANLEY TOMLINSON K.C.M.G., LL.D

WILLIAM TALBOT'S book is an engaging blend of a romantic story with an autobiographical account of life in a remote country as it was lived more than half a century ago. The lovely island of Ceylon – as I still prefer to call it – was occupied, first, for 150 years by the Portuguese, then for 150 years by the Dutch and finally for 150 years by the British; and there are more Portuguese names, particularly Fernando, de Soysa and de Silva than there are names of an Asian origin. All this has given to the country a unique flavour and atmosphere of its own. I was British High Commissioner in Colombo from 1966–1969 and these were three happy years in which I made some lifelong friends among the urbane and highly civilised members of the educated intellectual minority of the population.

It was in the mid-nineteenth century that, after an abortive experiment with coffee, British settlers began to grow tea: and when I arrived in Colombo in 1966 there were 150 British tea planters, most of them living in beautiful bungalows at heights of four or five thousand feet above sea level.

Then there was a major change in the political background. The determination to achieve total national independence in accordance with the *Panch Shilah*, the five principles enunciated by the Indians, became an irresistable tide. The British Government of the day decided wisely not to be Canute but to co-operate with the new nationalism. It was this nationalism that led the Sinhalese Government to nationalise all the British-owned tea estates; and I was asked by the leading tea companies to go to Colombo and try to secure a just settlement of the problem. The negotiations began inauspiciously with the Sinhalese attitude totally *non possumus*. In the course of two long visits, I urged the Sinhalese negotiator – an amiable young

Marxist Professor of Economics from Kandy University – to comply with international law on this matter, which was that a government seizing property from foreign owners must pay compensation that was "prompt, adequate and effective", and I also said that British tea companies had spent more than £2,000 on every acre of the vast acreage under tea cultivation to create the infrastructure necessary for the development of a viable tea industry. My efforts met with some success.

I have written too much already. But I will end by expressing the hope that some of the readers of William Talbot's book will take a holiday in one of the many splendid holiday resorts which now exist in the area about which he has written so interestingly.

INTRODUCTION

THE P & O Branch Liner S.S. *Barrabool* ploughed her way sedately through the Red Sea.

The afternoon had been hot – swelteringly hot. Most of the passengers had spent the time on their bunks. Now, in the cool of the evening, young William had found a corner of the boat deck where there was a breeze from the way of the ship. Not yet nineteen, he was on his way to Ceylon to be articled to a distant cousin as a 'creeper' on a tea plantation.

Captain Leicester Green had agreed, without much enthusiasm, to take William – at, of course, a suitable remuneration – as a pupil for six months. He should use the time on the voyage to study Wells' *Coolie Tamil*. William was doing his best with the vocabulary and the grammar, but his mind wandered.

He remembered the solemnity of the leaving service at Christ's Hospital and the Headmaster's final words: 'I charge you never to forget the great benefits that you have received at this place,' before he marched down the aisle of the chapel for the last time. His joy when he learnt that he had succeeded with his London Matriculation, only to be followed by the realisation that his war-widowed mother could not afford to send him to University. Then the paucity of opportunity for impecunious school-leavers in England during the 1930s. Now in May 1933 he faced a new world – it was, indeed, a great adventure!

Breaking free of his reveries, William returned to his book:

I am	–	*Nan irukren*
You are	–	*Ni irukruthu*
He is	–	. . . .

I

A CREEPER IN THE JUNGLE

Captain Green met William on the quay at Colombo, and was quite welcoming. He had brought his houseboy with him, so, while the boy collected William's luggage from the ship and loaded it in the car, they both went off to the Galle Face Hotel for coffee. As they entered the large reception hall, with its marble floor and electric fans overhead, a boy in a spotless white coat came to greet them. 'Salaam Master,' he said, putting the palms of his two hands together, and giving a slight bow.

'Coffee,' said Leicester with a dismissive gesture.

William had time to look around him – this was his first sight of one of the well-known oriental hotels. He found it impressive, almost too grand. Interpreting his thoughts, Leicester said, 'This is one of the meeting places in Colombo; the British are proud to show off in a place like this, but later on it will all come down to reality.'

The coffee arrived. 'What's that comb doing in the boy's hair? It looks like tortoiseshell.'

'Yes, it is, he wears it to show that he is a high-caste Sinhalese, and that he does not carry luggage on his head.'

The break for coffee did not last long. Velu, the houseboy, had loaded William's tin trunk upon the rear luggage rack of the Morris Ten hooded car, and was waiting in the road outside. They set off on the three-hour drive for Leicester's tea estate: Kadien Lena, Kotmalee, Kelnie Valley. The journey made quite an impression on William. As they went through the outskirts of Colombo the streets were crowded with Sinhalese and Tamils, the women with tightly fitting white bodices and flowing loincloths, some of the more affluent wearing brightly coloured saris and long, draping headcloths. Often the women had a small child slung on one hip, but even so, managed to move with a great deal of grace. The men were mostly bare up top with loincloths reaching below their knees, or tied between their legs. Everything was very casual, with little regard for the rule of the road. A bullock cart would be travelling along the middle of the road, requiring a long blast on the horn. The driver would look back in surprise at seeing a car behind him, before moving to the side of the road – hopefully to the left! The local buses were crammed tight with passengers, some sitting on the roof, some hanging on the side. The drivers seemed to expect everyone to get out of their way, driving perpetually on their Klaxon horns.

One of the greatest impressions was the smell. Everywhere was the odour of wood smoke as fires prepared the daily curry. The roads were covered with dung from the bullocks and horses, and there was the arid smell of betel nut being chewed to create the red saliva which was then spat out onto the ground. William was to become used to this 'cocktail' of smells, but, for now, it was all so new and exciting.

A quick stop to buy the standard pith topee (which was much cheaper than one would have been in London) then they turned off the main Kandy road to start the 1,500-foot climb up to Kadien Lena. The sun was setting now and the road was narrow. All Leicester's concentration, and indeed energy, were needed to negotiate the many hairpins on the estate road. The engine of the car whined ever higher as it thrust forward, the tyres screeching on the loose stone on the corners. Darker now, the lights of the car picked up the close canopy of the trees above the road while, looking down, William could see the outline of the Kelnie river far below. At last, turning a bend they came suddenly upon the estate coolie lines. The day's work done, many of the men were squatting on their hunkers chatting and waiting while their women cooked the evening curry. The same smell of wood smoke was everywhere. A few of the men got up and acknowledged Leicester as they drove past. A pariah dog ran along beside the car, barking furiously, and a small child that had been playing in the road ran to its mother.

Two more hairpins and another two hundred feet and the Bungalow came in sight. The electricity, generated from a nearby stream, bathed the building in a welcoming blaze of light, and as they climbed the steps up to the verandah, Veeraswamy the Cook Appu appeared to greet them. He spoke remarkably good English, in quite a cultured voice, and was quick to enquire after Leicester's health and whether he had had a good journey. He turned to William as if to sum him up, before breaking into a broad, white-toothed smile. Then he was gone to attend to the needs of his master.

Velu showed William to his room, which was simply furnished. There was an English bed with two sheets and a blanket draped below a mosquito net. At three thousand feet one did not need much cover at night. There was a cane chair and a small table by the bed, a basin in the corner standing on a teak wash-hand stand and an almirah with a mirror. William's tin trunk was lying on the floor which was covered with coconut matting. Round the corner was the lavatory, containing what was called locally the 'thunderbox'. There was an earth bucket and shovel. William learned later that the thunderbox was emptied daily from an outside access by a low-caste coolie called the *Vasa Kuthi*.

On returning to his room William unpacked his tin trunk, putting a photo of Dolly, his mother, on the small table by his bed. When he had finished, he went out into the Bungalow and found Leicester lying back in a long cane easy chair with a whisky and soda beside him. He beckoned William over but did not offer him a drink, saying rather pointedly that if he had not got his own supply it could be bought in the village. William had never tasted the stuff, nor had he any desire to, but in a flash of better judgement, he did not respond.

Veeraswamy came to announce dinner. They sat at a large round table by candlelight. William could not remember what they had – it was all so strange and unreal. There were geckoes crawling about on the ceiling, frogs croaking outside and fireflies flitting about in the verandah. They were two white men, miles from anywhere in a foreign land.

During dinner Leicester relaxed a little. He enquired about Dolly, talked about his own parents and said that Diana, his wife, was in England, having taken their son Richard to school. Though he did not say so, he gave the impression that Diana had been away quite a long time and that he missed her and his son. As things turned out, it might have been easier for William, now still only eighteen, had she been there. Then Leicester mentioned the 'boys' – Veeraswamy had been with him for ten years and was utterly reliable. He did the housekeeping and cooking as well as looking after Leicester's clothes. Velu, the houseboy, had only been with him for two years, but seemed bright enough. He would probably move on to being someone's appu in a year or so. He was learning to speak quite good English and would generally attend to William's needs.

After dinner they went out onto the verandah and Veeraswamy brought them coffee. When he had gone, Leicester's attitude seemed to change.

'Tomorrow, and as long as you are here, you will call me 'sir'. We shall change into dinner jackets each night to keep up appearances before the 'boys'. You will be ready for dinner at 8.00. Tomorrow morning, I have arranged with Velu to call you at 5.30 a.m. Muthiah, the Head Kangany, will be here at 5.50 to take you to the muster of the coolies at 6 a.m. You are to attend muster on every working day, which will exclude Sundays, except during the 'rush' season when there will be work every day. At muster, Muthiah, who is in charge of all the three hundred and fifty coolies here, will allocate work for the day. He is responsible to me for their discipline. He speaks no English and has been instructed by me on no account to attempt to speak to you in that language. It is your job to learn his language quickly – by the time you leave here you must be able to speak Tamil sufficiently fluently to manage your own division. You are to stay with Muthiah during working hours for the first week. After that I shall allocate jobs to you on your own. After muster you can return to the Bungalow for breakfast at 7 a.m. but you are to be back in the fields by 8.00. Lunch, which we call 'tiffin', will be at 1 p.m. and, again, you are to be out from 2.00 to 4.30 when the coolies have their leaf weighed in, and then men are credited with their work. Later you are to keep the checkroll, but for now this is done by one of the older kanganies, under the supervision of the Teamaker.'

Whoof! thought William, and I only arrived in Colombo this morning! But he restrained any response.

Veeraswamy arrived silently, barefooted, to take Leicester's empty whisky glass. 'Master been in Ceylon before?' he said to William.

'No,' he replied, 'this is my first day. It is all very strange!'

'We are all very happy here,' said Veeraswamy. 'We hope that master will be, too.'

'Thank you,' replied William. 'Velu will call me in the morning?'

'That has been arranged, master.' With that, he glided away.

By now Leicester was reading a book. 'Good night, sir,' William said, thankfully taking his leave. It had been quite a day, he thought, as he crawled under the mosquito net . . .

The next thing that William knew was the sound of Velu hammering on the door with a cup of tea. 'Good morning, master,' he said cheerfully as he drew the curtains. William could hear the drums beating furiously down in the lines to summon the coolies. Jumping out of bed, he put on the new khaki shorts, shirt and stockings that had been bought at Lilleywhites. Drinking the tea as he donned the topee that he had bought the day before, he walked outside to find Muthiah already waiting for him. William felt slightly awkward in his 'get-up'. It was all so new and stiff – he must have looked a sight. This time he initiated the 'Salaam, Muthiah,' which seemed to please the Head Kangany. He was an imposing-looking man wearing a clean white turban wound round his head, with a loose end hanging behind which swayed when he moved his head. He was clad in a white coat and loincloth, or *veti*, and was barefooted like the others. He was younger than William had anticipated – about forty, with a full face and white teeth, tinged with red betel when he smiled.

'Salaam, Sinna Dore,' he said. '*Kuda vara*,' (Please come with me). Muthiah led William down to the lines where all the coolies were assembled. The women wore quite gaily coloured bodices covered by a sari which was also covered, below the waist, by a thick blanket (*kumbly*) to protect them from the wet tea bushes. The men were mostly either bare up top or with a simple shirt above their loincloths. Those who were to pluck the tea had wicker baskets slung from their heads by cords. The men, who did the field work, carried pruning knives, forks or mamoties (a kind of spade with the blade reversed so the earth is pulled). After dividing them into various groups, they were sent off to their tasks in the charge of a kangany, to whom Muthiah gave what appeared to be instructions. All this took about half an hour after which he pointed to eight o'clock on his watch, saying, '*Kalame tini*,' (Breakfast) and politely left.

Returning to the Bungalow William went to his room to shave before going in to breakfast. There was no sign of Leicester, so he sat down. Veeraswamy brought him cereals with boiled milk, an egg with toast, and a large slice of a yellow fruit like melon, but called papaya. William shuddered as he tasted it – more like the smell of turpentine than a fruit.

'Eat it, master, it is good for you!' said Veeraswamy. No doubt he would get used to it – it was the local laxative!

The week that William was to have in company with Muthiah passed quickly enough. However brutal Leicester may have been in his instructions, there was no doubt that he learnt more from the Head Kangany in that time than he would have done in a month of Sundays from Leicester himself. As for the language, he was, of course, learning by ear and not by the written word. He would have to rely on Wells for that, but the fact that he was surrounded by the spoken word, with no retreat into English, worked wonders.

The week was not without incident: one day they were in a 'field' where the coolies were weeding with hand scrapers round the base of the bushes. William, finding what appeared to be a small tree that had been left, called Muthiah over and made to pull it up. '*Ille! Ille!*' shouted Muthiah. '*Antha vaga maram irukirathu!*' It turned out that the 'weed' was a little Albizzia tree specially planted for shade cover and green manuring! On another, and perhaps more serious incident, William was trying to engage, in his halting Tamil, the attention of one of the kanganies. This individual, looking rather sceptically at the Sinna Dore, said, '*Velank' ille*,' (Do not understand) and moved away. Muthiah was onto him in a flash, telling him to come back, and together they sorted out what William was trying to say. Later, William and the kangany became good friends.

As William followed Muthiah around the estate, the daily pattern of work began to emerge. Much depended upon the allocations at the early muster; the younger women were sent to the newly pruned fields where the leaf was growing fastest. The pluckers were paid by weight of leaf, and it was important that the breadwinners of the family should be given the best opportunities. The older and pregnant women went to the fields which were near the end of the pruning cycle, and where the leaf was smaller. Here, the 'plucking table' was higher and needed breaking back more often, so that progress was slower. On arrival, Muthiah would ask the kangany in charge how many coolies had arrived, and check that they were at the correct locations. There was a tendency to try and present themselves at the better fields.

When they had been picking for about three hours, the pluckers were called down onto the path, where they undid the *kumbly* or blanket around their waist before laying it upon the ground before them. Then, having squatted on one corner, they emptied the wicker basket in a heap on the *kumbly* and began to pick it over to extract the coarse leaf and stalks, which they discarded. This was a time for some relaxation, and there was much chatter and banter among the women, spoken so quickly that William gave up hope of trying to understand. When the leaf was ready for weighing, the kangany hung a large wicker weighing basket from a Salter spring balance. After weighing, the leaf was thrown onto a tarpaulin and gathered up into large hessian sacks, which were later carried down to the tea factory by some of the men, who nimbly hoisted a fifty-pound sack onto their heads.

Then to the pruning field. These were the ablest and strongest men on the estate. It was obvious that Muthiah had respect for them, though he would allow no nonsense. Their job was to prune the head off the tea bushes leaving a bare framework of branches cut to a low, spreading table. They used a single-bladed and slightly curved knife honed to razor sharpness. Each had his own knife which he tended with loving care. With the knife in his right hand he would bend the branch outwards and downwards, making a short neat cut. To bend the branch too far would result in a split stem; not to bend it far enough would mean a long sloping cut which would be equally disastrous. The coolies were given the task of pruning about one hundred trees each per day, including collecting and burning the prunings. With 2,500 bushes per acre, this would take a gang of ten men about two and a half days to prune an acre.

William asked to be allowed to prune a bush and promptly split the branch. This then had to be taken out at the base. When he had finished the bush, Muthiah gave him cautious approval. However, when William returned later he found that 'his' bush had been tactfully 'cleaned up'!

The evening muster was a highly organised affair. The coolies left the fields at 4.30 p.m. making their way down to the muster ground near the tea factory. Muthiah sat at a large kitchen-type table with the checkroll in front of him. Someone hurriedly found a chair for William. The pluckers picked over their leaf while the men filed past the Head Kangany in order of their gangs – the pruners, with their knives neatly tucked into their belts, were first, each giving his name for Muthiah to record his day's work. The kangany in charge of each gang nodded his approval as each name was given. Muthiah, of course, knew their names, but it was a matter of discipline that the coolie should ask for his name to be recorded.

It was nearly 7 p.m. by the time William returned to the Bungalow. He had to hurry over his bath to be ready, in his dinner jacket, at 8 o'clock. Velu had laid out a clean vest and pants – a little piece of civilization for which William was duly thankful. He had been so busy keeping up with Muthiah and listening, and trying to follow what was going on, that he had not noticed how hot and sweaty he had become. When he came out onto the verandah Leicester was already reclining in his long chair with both legs resting on the long supports before him. William noted that his pipe was not drawing well and that the saliva was dribbling down his chin. He looked up from his book, saying, 'Well, how did you get on today?'

William had anticipated such a question, and had turned over in his mind some of the questions that he wanted to ask Leicester. However, something in Leicester's attitude, and the fact that the matters probably needed further research, made him reserve them for the time being.

'Very well, thank you, sir. Muthiah is doing his best. The language is the most important thing at the moment.'

'Yes,' said Leicester. 'You will pick it up soon enough, as I had to when I left India.'

Veeraswamy appeared to announce that dinner was ready and the two men sat down to candles flickering on the table. It was already dark outside, but William did not notice the difference from the long summer evenings in England. During dinner Leicester suddenly turned to him, saying, 'You play rugger, don't you?'

'Yes,' said William with a touch of enthusiasm in his voice.

'There is a rugger club in the Kelney Valley – where do you play on the field?'

'Scrum half,' was the reply.

'There is no play until after the monsoon in August, but I will take you down to the club sometime so that you can have a word with the Secretary.' With that, Leicester returned to the contemplation of his soup.

William was thankful to retire to his room after dinner, so that he could write to Dolly. He hoped that he would soon hear from her.

The first week passed quickly. Muthiah showed William all the work that was going on. Coolies were weeding under the newly pruned bushes with a small hand-held scraper so that they could get close to the base of the bush. An exhausting and back-breaking job that demanded a great deal of stooping. This was done by most of the older men and some of the younger women with small babies literally hanging from their breasts. Then there were gangs of coolies manuring. The bushes were over-planted with Dadap and Albizzia which were being cut back to provide green manure. Boys with small sacks on their backs ran along the rows shaking a small tin (*sundo*), placing a measure of inorganic manure to each tree, later to be forked in with the green manure.

Most of the rows of bushes were planted in fields with steeply falling contours, resulting in a heavy run-off of soil during the monsoon rains. In an effort to prevent the soil erosion, ditches were cut around the hillside, to a slightly sloping gradient, to catch the eroding earth. These ditches soon became silted up. Re-cutting the drains was another not very popular job. William noticed a flurry of activity as they arrived, from which he assumed that this was not normal practice.

On a day near the end of the week, Leicester arrived in the pruning field to give William his first opportunity to see how he dealt with the coolies. Things went well enough: Leicester knew most of their names, asked after their families and whether they had any complaints. He discussed with Muthiah and the kangany in charge the future work programme and, after a quick word with William, departed. Perhaps he was embarrassed by the latter's presence, but William was surprised at Leicester's lack of fluency in Tamil; there were many English words mixed up in what he said – so much for his exhortations! Nevertheless, William was determined to master the language as quickly as possible.

The next week, as had been promised, William was given charge of a gang of four coolies, whose task was to plant the seed of a large perennial type of vetch called by the Tamil name of 'killi-killipu' because when the seed pods were dry they rattled. The idea was to plant the seed flat along the contour so that, when the plants grew, they would hold up the soil erosion – a good idea, thought William. William had a dumpy level (a simple form of theodolite that worked on the pull of gravity) by which he could check the contour levels. There were also three painted staves for sighting. Having levelled the dumpy, the coolies were sent out along the line, beckoning them to move up or down to the correct line. A young boy carrying a small sack of seed on his back would then sow and cover the seed with a weeding scraper.

William was glad to have a definite job to do; by now he could talk, haltingly, to the men, who ceased to be 'just coolies' and became human beings, with their own families and background. Ramasami, who carried the dumpy for William, was a young man who had come over from southern India four years ago, but who still kept in touch with his parents there, sending them money on occasions. He was now married with an eighteen-month-old son. His wife was in the plucking field. He hoped to be a pruner before long. Vengadasami was older; William guessed about fifty, and rather dour. He was obedient without being enthusiastic. On the other hand, Kuppai was a young lad of eighteen who was anxious to show his agility and willingness. His parents were both on the estate where they had lived for many years and were obviously part of the backbone of the labour force. Amasi was just fifteen when he started to work on the estate. The other coolies lost no opportunity to tweak him as he went along the line with the seed. It was either '*Suruka, suruka, Amasi,*' (Quickly, quickly Amasi) or '*Ille, ille Amasi, angatu, angatu!*' (No, no, over there!). The boy took it all in good part, so William saw no harm in it, but sometimes their instructions were not correct, so that he had to intervene, particularly before Amasi had sown the seed!

When all the killi-killipu seed had been planted, William reported to Leicester that the job was finished. 'Right; we shall see how well you have used the dumpy when the seed comes up,' was his somewhat wry reply.

William now began to take charge, with the kanganies, of the various gangs on the estate. Soon, he began to know their names, and those of their relatives, and found this a most useful subject upon which to get them to talk to him. Although they were forthcoming enough when he got to know them, there was an innate shyness, particularly among the women, who used to cover their mouth with their headdress when he spoke to them. Some gave the impression that they did not really expect to be spoken to by a white man, and were embarrassed by it. Supervision of the work itself was not very difficult, and depended upon the character and capacity of the kangany. William found that it was better to work through them, rather than upbraid individual coolies for their work. The kanganies were mainly appointed by the Head Kangany, and were often relatives. This meant that they were of the same caste, usually a high one, and that the coolies were respecting them on this account as much as for their powers of discipline. William found later that, historically, many of the coolies had been brought over from India in quite large numbers, after being recruited by the Head Kanganies, who had paid some of their expenses. In a sense, therefore, they were 'owned' by them. William pondered that, while clearly the white man was 'boss', the Head Kanganies were good people to keep on the right side!

As he got to know more about them, William tried to select certain men for certain jobs, according to their capacity and aptitude. In this he met something of a barrier from Muthiah. At the time, William thought that this was something to do with the Head Kangany's desire to preserve his own authority. Later, he concluded that, in addition, the caste and family preference system was probably as important. This also proved to be a valuable lesson for the future.

He was now receiving a weekly letter from Dolly for which he waited eagerly. Perhaps some of his letters sounded rather homesick: he could hardly have entered into a lifestyle in greater contrast. Leicester he referred to as The PTB (The Power That Be). It was not until later that he realised his good fortune in the grounding that he had at Kadien Lena. Dolly herself was keeping well enough. There were messages for William from the MacDermotts, the Sweets and the Gaze girls and the Christ's Hospital Old Blues Club had asked for his address.

June turned to July. William was still on his edict to remain in the fields all day, but in the days before the monsoon it was becoming progressively hotter and at times was almost stifling. He was sweating copiously, and he found that sweat glands in his armpits and crutch were becoming sore, the latter being almost raw. In the end, almost apologetically, he mentioned this to Leicester, who showed no sympathy. 'Wash more and eat more salt,' he said. William took this to mean that he was at least allowed back into the Bungalow during the day for a bath, which he did.

Then the monsoon broke. How thankful William was – he stood out in it feeling the rain running down his body with its blissful cooling and healing property, thanking God for such mercies. There was no question of any cover – when the sun came out his clothes dried upon him – but it was comparatively cooler!

By mid-August the monsoon had blown itself out. Keeping his promise, Leicester took William down to the Kelney Valley Club. William reflected that, apart from a few shopping trips, this was the only social contact that he had had with the outside world since his arrival two and a half months previously. Without his wife Leicester was no club man, and without transport there was little that William could do. Besides, by Leicester's edict, he was fully engaged on the estate!

William was introduced as 'the Creeper from Kadien Lena'. Not a very complimentary beginning, but he took an instant liking to Gordon Lacey, the Club Secretary and captain of the team. Gordon, a man in his late twenties, had played for Eastbourne College. He had played at 'Housey'.[1] Yes, they needed a scrum half for their game against Dimbula on 28 August. Guy Simpson, who was an SD on Ampitiya Kandy, further up the valley, would pick William up at the entrance to the Kadien Lena cart road at midday and would bring him back. They would probably be late, so William should bring a good torch as well as his rugger things. William was delighted after the closeness and unreality of his existence: it was good to meet some kindred spirits again.

William had kept himself fit, but in the days before the match he took to running along the estate roads, hills and all, in the evenings before dinner. This made him late, but seemingly neither Leicester or Veeraswamy minded. The day came. As William set off down the cart road to meet Simpson, Leicester actually looked up from his seat on the verandah and waved.

'Good luck,' he said. 'Don't make a noise when you return.'

'Thank you, sir,' said William.

It was like walking into another world. Now just nineteen, William had

---

[1] Christ's Hospital

no idea that there were other planters who also had creepers. Although the youngest, he found that there were other estates with perhaps two or three SDs as well as a creeper on the same estate. They did not appear to be on the tight rein that Leicester imposed. Guy was due for his first leave in December and was keen to hear about events at Home. The journey to the Club passed quickly.

From the kick-off, William was, again, in his element. Although he was not without diffidence for his team-mates, the game offered an opportunity to work off some of the restraint and self-effacement that had been imposed upon him during the recent months. He was now doing something that he really knew about! Younger and fitter than his opposing scrum half, he was regularly round spoiling the other's passes from the base of the scrum. William had an eye for the referee who, perhaps from inexperience, managed to get himself on the wrong side of the scrum to see William's tactics. Unfair? Yes, thought William, but all is fair until it is checked! Once he actually took the ball from the scrum half's fumbled ball. Half-expecting the whistle, he flung it out to his own fly half as the game continued.

The Kelnie stand-off half was an experienced player after William's heart. They soon established a rapport which anticipated the other's intentions. Within a few minutes of the start, from a scrum in midfield, the Kelnie hooker took a ball against the head. From a long, clean pass from William the fly half dropped a perfect goal – four points up! There followed two converted tries by Kelnie, the last being the result of a decision by William to hold the ball in the scrum near the Dimbula line, and the forwards carrying it over for a touchdown. Dimbula scored one converted try leaving Kelnie the winners by 15–5. A most worthy win!

'Well done, William!' said Gordon Lacey as they came off the field. 'You have certainly been taught some aggressive rugby! It is the first time that we have beaten Dimbula for five years, and last year they were top of the Provinces League. By the way – do we have to call you William? Wouldn't Bill be better?'

Outside Kadien Lena, William had grown up – from now on it was to be 'Bill'.

It was past midnight when Guy Simpson dropped Bill at the entrance to Kadien Lena cart road. During his three-mile, 1,500-foot climb up to the Bungalow, he had plenty of time to reflect upon the happenings of the evening. It had been a boisterous affair with much singing, dancing and ribaldry. Bill did not drink – he had no money to buy and was reluctant to accept a round which he could not repay. It was a change to see some English women about again. Most were rather middle-aged and to Bill's senses rather sophisticated. Most of the talk was about the children at home, their next leave and their last bridge party. Seeing Bill a little out of things, one rather portly lady came up and insisted that he should dance with her. Dancing was not one of his best accomplishments, but he was grateful for her kind thought, and she was glad to hear about England. On the whole, he was glad when Guy suggested that it was time to go. He had enjoyed the whole day, including the evening, and realised that he must overcome his shyness.

By now he was most of the way up the mountain. The noises in the darkness of the jungle were at times quite eerie, the most noticeable being the sound of rushing water, of which there was still a great deal about after the torrential rains of the monsoon. Bill was now wearing his mosquito boots – leather and canvas long boots strapped below the knee which kept the water out as he clambered over the Irish drains let into the road. He was thankful for his torch and would have to remember to get some new batteries before the next match.

Veeraswamy had left the back door open. He was glad to climb into bed. He had about three hours for sleep before Velu called him for Sunday muster. Ough!

The match against Badulla was 'away' over the other side of the Island, via Newara Eliya. This meant that Guy would be picking Bill up soon after breakfast. Surprisingly, Leicester did not demur; he probably felt, rightly, that William's chances of a job depended as much on the social contacts, through his rugger, as on any business contact that he could give.

In the field, Bill was now becoming reasonably fluent in the coolie Tamil as it is used on the tea estates. He was becoming more of a Sinna Dore whom the coolies could respect. He had some ideas of his own for saving time and work. They respected him for it. Sometimes Muthiah actually expressed his thanks for his suggestions!

The coolies were paid every month according to the 'names' and credits that they had in the checkroll. The clerk in the office prepared the paysheet, and Leicester would take it down to the bank, taking a trusted bodyguard with him, to draw the money. Upon his return he would deposit it in the office safe until the coolies were mustered for pay. The October pay day fell on a date when Leicester had to attend a director's meeting in Colombo, so he collected the money the day before and told Bill to pay it out.

All went well until the end when Bill discovered, to his horror, that he was about fifty rupees short on the payroll. He checked the roll with the clerk, but could find nothing wrong. On Leicester's return he had to admit, ruefully, that he was short. Leicester appeared to be furious. 'Well, what are you going to do about it?' he demanded.

By now Bill felt that he had already apologised enough, and pointed out that he had no money with which to make good the deficiency. Whereupon Leicester grudgingly agreed to replace the shortfall. 'Did you check the money before you started to pay?' asked Leicester. To Bill's further embarrassment, he had to admit that he had not. Lesson learnt.

It was not long before Bill learnt that Leicester had deliberately removed Rs. 50 from the pay bag, to teach Bill to count the money beforehand. Bill thought that there might be other ways to bring the lesson home!

September to October. One morning at breakfast Leicester asked, 'May I know how long you propose to stay here?'

By now, Bill had become used to these rather pointed questions. 'I rather think that my mother is relying upon you to let her know when you suggest that I should try for a job,' he replied. 'Did you not mention a period of six months, but that you could not guarantee employment?'

'Yes,' was the reply, 'and the six months will be up on 25 November. You should be thinking of moving on.'

'Yes, of course,' said Bill. 'Can you suggest what I should do, please?'

'Advertise in the *Ceylon Times*.'

Bill sensed that that closed the conversation, so next evening he studied the advertisements in the paper. He then drafted:

'Public Schoolboy, 19, at present creeping on Kadien Lena, Kotmalee, seeks first appointment on a tea estate. Please contact W. T. Baker c/o Captain E. L. Green at the above address. (Tel. Kotmalee 785.)'

'Yes, that will do,' said Leicester.

A week later there was a letter for Bill. His heart beat a little faster as he opened it. It was from a Mr E. L. Doudney, Managing Director Messrs. Liptons (Ceylon) Ltd. and was short and to the point.

'With reference to your advertisement in the *Ceylon Times*, this Company may have a vacancy on one of its tea estates. Please attend our offices at Galle Face, at twelve noon on 12 November, for interview. Kindly confirm.'

Bill immediately wrote off the confirmation, and when the day came Leicester drove him down to the station. He found a taxi to take him to

Galle Face to arrive in good time. He wore the grey flannel suit that he had brought out from England for such occasions – rather hot, but bearable provided one did not move around too much!

Mr Doudney was a large man, fair and balding. He asked Bill about Kadien Lena and what he had been doing there, a question that gave Bill ample opportunity to express himself. The fact that he was Captain Green's cousin seemed to help. After they had talked for about five minutes, Mr Doudney said:

'The appointment is on our Dambatenne Group at Haputele. You would be an assistant superintendent on an isolated Division where there is no cart road. Are you prepared to accept that?'

'Yes,' said Bill, without hesitation.

'Your salary would be Rs. 250 per month, terminable by three months' notice on either side.'

'That is quite acceptable to me, sir,' Bill said.

'Right, you can start on 25 November. Please report to Mr Gordon Prior on that date.'

Bill rose to go, but Mr Doudney restrained him. In a less formal voice he said, 'You have just come out from England – how are you placed?'

Bill took this to mean financially and started, 'Well, I have £96 in the bank here . . .'

'No, I mean for household utensils and so forth.'

'No, I'm afraid that I have none,' replied Bill.

'Well, we will see what we can do.'

Bill left with a glow of satisfaction and much anticipation.

Leicester seemed pleased that Bill had found himself a job. He didn't say whether he had been asked for a reference, but there was little doubt that he had. Leicester arranged a late, and rather hurried, introduction to the tea factory, and a tour with Mr Patel. It was obvious that he regarded the duties of a creeper to be in the field, and not in the factory. Bill resolved to learn more about tea making when he got to Dambatenne.

Bill wrote to Mr Gordon Prior to introduce himself, and to give details of his arrival. Later, Leicester, who knew Prior as 'Paddy', received a call from him. He would pick Baker up at Haputele Station on 25th. Bill also wrote an excited letter to Dolly and had a cable back from her: 'Delighted – proud of you. Dolly.'

Bill went to his last morning muster on 24 November to tell everyone that he was leaving. He had an especial nod for the kanganies, putting the palm of his hands together, in Hindu fashion (he was not sure that Leicester would approve). Finally, Bill made a little speech in his best Tamil, which he had prepared and rehearsed beforehand, thanking Muthiah for all his help and patience.

When Bill had finished, Muthiah turned to him, his dark eyes above a beaming grin. Then, in perfect English, he replied: 'Thank you, master, for what you have said. It has been a great pleasure to have you here, and we wish you every success at Dambatenne.'

Bill looked at him in astonishment. 'You old rascal!' he exclaimed. 'Then you spoke English all the time that I was struggling!'

'Yes, master,' replied Muthiah, 'the Peria Dore wanted it that way – it was for the best.' And there was little doubt that it was. Bill shook hands with Muthiah and they parted.

Next day, after goodbyes to Veeraswamy and Velu, Leicester drove Bill down to Kotmalee station. On the way down he said, 'I assume that you are on probation?'

'Yes,' said Bill. 'On three months' notice on either side.'

'Well, if your probation is not confirmed, you must not expect to return here.'

This was the final taunt of a rather unhappy man, but Bill let it pass. He was determined to have his appointment confirmed.

Bill had much to thank Leicester for; indeed, his offer to take him as a creeper started him off on his career. Whether it need have been quite so traumatic is another matter!

2

GREAT EXPECTATIONS

Monty Villiers, the Senior Assistant at Dambatenne, was there to meet Bill on his arrival. He had a pleasant easy manner, and explained that Paddy Prior had asked him to find a bed for Bill on the first night. His wife Joanne was in England, having taken his two children to school, but his boy Muttu would look after him.

Paddy would see him tomorrow, before taking him up to Mousakellie, which was to be his Division. There were three Divisions; the principal being Dambatenne Division which was near the factory. Then there was Bandara Eliya Division, at the top of the estate, at some six thousand feet. It made some of the best tea in Ceylon. Finally, Mousakellie was on the hillside ranging from four thousand five hundred feet to the Bandara Eliya boundary. His brother, George Villiers, was moving from Mousakellie to the upper Division.

As they drove up to the factory at Dambatenne Bill realised that this was on a much bigger scale than his previous estate. There was a deal of bustle on the forecourt as huge overhead ropeways brought leaf down from the outlying Divisions, to be briskly unhooked and carried into the factory. Large lorries were taking on tea boxes for transport to Colombo. There was an air of efficiency about the place.

Paddy Prior was a very different character from Leicester. An RAF pilot during the First World War, he, like most Irishmen, had a strong sense of humour and understanding. He felt fortunate at having survived the war and was glad to help the son of one who hadn't. He listened to Bill's account of what he had been doing at Kadien Lena and said that he knew Captain Green and was sure that he had given a good grounding. Mousakellie was an isolated Division, but most important in the grade of leaf that the estate produced. The management of labour was the all-important factor. New labour was not easy to recruit in that situation. The Head Kangany, Kadi-raveil, had originally introduced much of the labour force of seven hundred coolies. Baker would find Kadiraveil co-operative to a point, but inclined to get on his dignity. Baker should 'read the signs' and have good reason for pursuing something against his wishes, particularly at the beginning. He would find that many of the kanganies were related, and of the same caste as Kadiraveil. He was sure that Captain Green had taught him that the secret was to retain authority but be fair and understanding by being prepared to make justifiable exceptions. Bill remembered this good advice. Finally, Baker would be expected to make his own decisions, but he or Monty Villiers would be available if required.

So it was to be 'Baker' and 'sir' again – quite like being at school – but at least he knew where he was.

Suddenly Prior said, 'Do you ride a horse?'

'Yes,' said Bill, 'but not very well.' Actually, he had never ridden in his life, but saw no reason for saying so.

'Well your horse, Eros, is waiting with your horse-keeper. We will ride up to Mousakellie now, and I will introduce you to Kadiraveil. Your *saman* (luggage) will go up on the ropeway. I have already asked Kadiraveil to find someone to look after you in the Bungalow, until you can find a cook appu. I suggest that you advertise for one, preferably a boy who has been with a planting family. My wife Lydia asked our head boy to contact your beef box coolie to make a simple order on your behalf, and the coolie fetched it yesterday from Haputele Cold Stores, where we all get our supplies.

'You seem to have impressed Mr Doudney at your interview in Colombo; at any rate a box addressed to you which, I imagine, contains utensils, has also arrived and gone up on the ropeway. Normally our superintendents are required to provide everything except the basic furniture. You may find that you need other necessities – I suggest that you order them from Greenways in Colombo. They are very good, and with an address like this you will be able to get anything 'on tick'.'

Bill was surprised and rather flattered by the amount of preparation that had been made on his behalf, and said so.

With that, they went through into the Bungalow drawing-room to meet Prior's wife and young daughter Monica, where they sipped a freshly iced lime juice.

Bill found Eros amenable enough as he followed Prior's horse through the fields on the Dambatenne Division. On the way, Prior acknowledged, with a wave of the whip, the salaams of the many coolies they passed. Many got a word of encouragement by name; he seemed to be well liked and respected. Soon the road began to rise and ascended, by hairpins, towards a gap in the mountains. 'The Gap, called Badger's Patch, is the boundary between the two Divisions,' said Prior. 'The reverse face is your Division.'

On reaching the Gap, at about five thousand feet, there was a magnificent view on both sides which they stood for a moment to admire.

'Why Badger's Patch, sir?' asked Bill.

'Well, it's not a very nice story, really. Badger was one of the old original planters at the end of the last century. The story goes that he was returning one dark night, on his horse, and was so drunk that he fell off at the Gap, and never got up. Next morning he was found covered in leeches and quite dead!'

'Oh,' said Bill, with a slight shudder.

Before leaving the Gap, they looked back over Dambatenne. There before them lay the factory, the Big Bungalow, the coolie lines and all the activity in the fields – truly a panorama of that part of the estate. On the other side of the Gap the road descended, again by hairpins, down to the Mousakellie muster ground. The bush telegraph had advised the Head Kangany that the Peria Dore and the new Sinna Dore were coming. He was waiting to meet them, a small man, in his late fifties, with black penetrating eyes. He had a white turban loosely tied round his head, a none-too-clean white jacket and a loincloth tied between his legs, coolie-fashion. Unless one knew who he was, one would not give him a second look, thought Baker, but it just shows how careful one must be! Prior remained seated on his horse, asking the kangany where the work was. Katheraveil described this in detail, but did not offer to accompany them.

Then to the Bungalow. This was a timber, white-boarded building with a red-painted corrugated iron roof, built in the form of a rectangle, facing the open valley. In the centre a protrusion formed a part of the living-room which ran the width of the building; there were two bedrooms, one double, one single, on either side of the living-room, with a small office at the far end. Outside, a covered passageway gave access to the rooms, and to the bathroom at one end. An extension provided the kitchen, with a wood-burning cooker and an earth floor, on the side of which was the boy's room.

The interior was wattle, plastered and decorated, and there was a cedar-wood ceiling, below the corrugation. The furniture was simple: a wooden table with four chairs; two long chairs with leg rests, similar to those at Kadien Lena; and a further table and chair in one corner. On the floor was striped coir matting, and there was a brick-built fireplace – something that Bill had not seen elsewhere in Ceylon.

Prior took Baker round almost apologetically. To some, it would have been a hovel; to Bill it was a palace – his first home of his own!

Pausing in the main bedroom, Prior pointed to the wooden bed on which the blankets and pillow were neatly folded. 'You will have to provide your own linen,' he said. 'Greenways will help you.' In the kitchen were some rather dirty-looking utensils, and the whole reeked of wood smoke. No wonder that the English women kept out of the kitchen! There was, of course, no electricity.

Prior left Baker to unpack his belongings. In Doudney's crate there were some china plates, some rather odd knives, forks and spoons, a milk jug, cups and saucers and an old Victorian-looking teapot which he dubbed 'Doudney's pot' – no disrespect intended, of course! Bill resolved to write and thank him. They had been neatly packed to survive the journey.

While he was unpacking, the house coolie came in to say that Raman, the horse-keeper, had prepared the meal for Eros and would master please come and inspect. Feeling a little guilty that he had neglected Eros, he went out to the horse-keeper to see the bucket of oats and bran, nodding his approval. This turned out to be a daily ceremony. The SD was responsible for the health and care of his horse, and there must be no question of short rations. Bill kept the key of the foodstore and issued the horse food on a weekly basis, but it was necessary to see that there were no 'slips' twixt cup and lip'. The stable was a strongly built stone building with a squatting place for Raman – he lived in the lines.

One attractive feature was a stream which ran between the Bungalow and the stable. Rising in the mountain above, the spring water was fresh and clear, cascading in rivulets on its way down the valley. A small dam above collected water for the Bungalow, providing a good head of water.

Bill sat down by the telephone. To make a call one had to wind the handle of a box. A voice would answer in English, 'Yes master?'

'Will you please get me Greenways in Colombo.'

'Yes, master. Estate or private?'

Bill hesitated. 'Private,' he said.

Almost immediately he was through. To his order of three sheets and pillowcases and four face and bath towels, he gave his address as Mousakellie, Dambatenne. 'Oh yes, sir, they will be on the train to Haputele this evening.' It seemed that such an address was an Open Sesame to credit! They duly arrived the next day – could England beat that service?

He then made a call to the *Ceylon Times* for a cook appu, having carefully phrased the advert. When he had finished, the house coolie arrived with a pot of tea and some rather odd-looking oatmeal cakes, which Bill guessed that he had made himself. They would have been better with some butter!

As he was drinking, the tom-toms started beating below for muster. Grabbing his topee, he set off down the stone steps of the paths through the bushes to the muster ground. Raman would follow later, with Eros, for his ride back. He did no more than greet Kadiraveil again and be introduced to some of the kanganies. He watched the routine which was not so very different, but he was enthralled by the ropeway and to see the sacks of leaf being clipped onto it. It was so silent and powerful.

He enjoyed his ride back to the Bungalow on Eros; one got a better view of the tea from horseback, but he was also a comparatively easy ride.

That night the house coolie prepared him some curry and rice and there was some canned beer in the ice box. Later he curled up under the blankets.

Bill 'felt his way' through the first day at Mousakellie. George Villiers had always kept the checkroll, so he told the Head Kangany that he would keep it as from the 1st December, leaving Periaveil, the kangany who had been keeping it (with some difficulty) to continue until then.

After the evening muster Bill had his bath, which had been prepared for him by the house coolie. This was in itself some feat. In order to heat the water, a large kerosene oil drum had been erected, endwise, on stone supports. This had to be filled with cold water by a pipe from the dam. A fire was then lit under the drum some two hours before the Sinna Dore was likely to want it – a good houseboy would estimate this to a nicety! The SD had an allowance of one Albizzia tree per month for fuel for cooking and for heating the water. Hot water for washing in the morning was heated on the kitchen stove.

After he had completed dinner Bill was about to sit down to write to Dolly, when there was a knock at the french window at the front of the living-room. Puzzled at what this could be – it was now quite dark – he opened the door, and was confronted with an amazing and spectacular sight.

There stood Kadiraveil clad in his best shining white attire, this time his loincloth hanging down to his feet. Behind him, he was surrounded by coolies with their tom-toms and wind instruments, with women dancers in gaily coloured saris, with gold beads round their necks and heavy ornaments hanging from their earlobes. The whole scene was lit by flaming hand-held torches, made from sacking dipped in kerosene. They had crept up through the tea bushes without a sound. The surprise was complete.

Bill was too dumbfounded to say very much. The house coolie hurriedly produced two chairs, which he put in front of the french windows, and, after Bill had sat down, Kadiraveil joined him. A small boy, called in Tamil 'Podian', advanced and placed a garland of sweet-smelling frangipani flowers around Bill's neck. Kadiraveil then got up to make a short speech of welcome and say that they had prepared a small *tamasha* for him. As soon as he had finished speaking the tom-toms started a frenzied rhythm to which the dancers gyrated with ever increasing momentum. In time Bill began to feel quite mesmerised by the scene and the infectious beating of the tom-toms. Here he was in the centre of a foreign land indeed.

The music stopped as suddenly as it had begun. One of the kanganies stepped forward carrying a tray, again covered with flowers, but on the tray stood a bottle of English whisky and a bottle of Gordons gin!

As the kangany came out, the pleasure drained from Bill's face. He was now faced with an awful dilemma; no one had warned him that this might happen. The cost of the drink represented more than the kanganies' monthly wages (they drank a form of toddy distilled from coconut milk themselves). Would acceptance signify that his favours could be bought for drink? Would he be in some form of debt to them? On the other hand, by refusing, would he offend the Head Kangany on his first day – something that Paddy Prior wished him to avoid? Should he ring Prior for advice? To do so would seem to be lacking in the power of decision, both before Prior and the coolies.

All these thoughts went flashing through Bill's mind as the kangany advanced. Finally came the thought: 'To thine own self be true.' He was sure that he ought not to accept such gifts, but, in so doing, he should say why not.

Bill rose to his feet, in what was the greatest test of his Tamil education so far.

'It is a great honour that you have done me tonight. I give you my thanks for the welcome that you have given me, and I know that we shall work well and happily together. I have not been in Ceylon very long, but I am already beginning to feel at home here. I was told, before I came, that Mousakellie was the best Division on the estate: your performance here tonight just goes to prove it!

'I am also honoured that you should offer me these gifts of whisky and gin. Firstly, I must say that I am still young enough not to have become accustomed to strong drink. Secondly, you who work so hard and well in the fields will understand that one must earn the payment that you receive from the estate. Similarly, in the two days that I have been here I have done nothing for you that would justify my accepting these good presents.

'I much appreciate the warmth of your welcome, and know that you will understand that I must decline your offerings.'

'My salaams to you all.'

This little speech taxed Bill's hard-learnt Tamil education to the hilt. He had trouble with the word 'to earn' and had to use 'work for' instead (*veile seia*). But it seemed to have the desired effect as they left as silently as they came.

On the whole Bill found the evening pleasant and stimulating, and hoped that he had defused the situation over the presents. He mentioned it to Katheraveil the next day and the old man simply grinned – he was not offended. Bill often wondered the motives behind the drink offerings – was he being tested for gullibility, and what would have happened had he accepted? Still, these were idle thoughts.

The Bungalow at Mousakellie lay on the five-thousand-feet contour. Behind, the land rose sharply to six thousand feet on the Bandara Eliya boundary. Below, it dropped to about four thousand five hundred feet at the muster ground and the lines, falling again to the lower boundary which adjoined another Lipton estate called Monrakandy, where the elevation was four thousand feet. One needed a strong pair of legs to cover this type of terrain every day! He was thankful that he was provided with a horse. It was Bill's practice, when he set out for the fields in the morning, to ride Eros up to the high ground from where he could have a good view of the Division as a whole.

In days gone by, a small chalet had been built at this viewpoint. Stories had it that Sir Thomas Lipton used to enjoy bringing important guests to Dambatenne, where they were suitably entertained in the Big Bungalow. The next day he would arrange for them to be taken up to this viewpoint, either on horseback or, for those who preferred, by carrying-chair. They were further entertained at the chalet, before being taken to the edge of what was, at that point, a precipice. There, Sir Thomas, his thumbs in his braces, would say, 'There, ladies and gentlemen; as far as the eye can see belongs to me, Sir Thomas Lipton.' It was then desired, and suitably arranged, that the guests should clap.

However, on one occasion, a junior assistant had somehow infiltrated the party. Unfortunately, he interjected: 'Oh no, Sir Thomas. In front of us here is the Mousakellie Division of Dambatenne, beyond that is your estate of Monrakandy, but beyond that is all Crown Jungle.'

Sir Thomas is said to have wheeled upon the unfortunate lad. 'Come and see me in the morning.' The following week an advertisement appeared in the Ceylon Times under the 'Situations Required' column.

Perhaps discretion is as important as know-how!

From the high point, Bill would hand Eros over to the horse-keeper and walk down through the tea to each of the gangs, having a word of encouragement with each of the kanganies, asking what target (*kanak*) he had set, and seeing how many coolies were in the field. He made a mental note (and sometimes an actual one) of the number present so that he could compare with the claims for 'names' made on the checkroll. On the way down he would call in on the lines, having a word with the elderly or the infirm who were not working, mildly chaffing anyone who, from a casual glance, might have been doing so. The Tamils have a word '*vekkum*' meaning 'ashamed', which can be useful when employed in the right place!

During the second week after his arrival the phone rang one evening. It was the wife of a planter over at Nanu Oya, not far from Haputele. They were going on leave shortly and had seen Bill's advert for a cook appu. Their appu would be out of a job when they left – would Bill like him? He would be thoroughly recommended. Bill was delighted. His salary would be Rs. 25 per month and half a bushel of rice. They would pay his fare to Haputele, if Bill would pick him up there. So Banda arrived. A quietly efficient man in his early thirties, who spoke good English, he was to be a faithful servant for the rest of Bill's days in Ceylon. Banda soon took hold of things at Mousakellie, ordered the beef box, looked after Bill's clothes, cooked for him, answered the telephone, and worked in well with the horse-keeper and garden coolie. Bill was a lucky man.

Bill had been on the estate for about two months, but he still had no transport. George Villiers had taken him to the Haldumulla Club once to introduce him, but until he could get there himself, he did not feel like joining. Paddy Prior mentioned this on one of his weekly rounds – he did not think that it was a good thing for Bill to be stranded on the Division, without being able to have some socialising. However, he did not pursue the matter when Bill commented that he had no way of getting about.

One evening he read an advert in the paper. A firm in Colombo was selling BSA motorcycles for Rs. 845 (£65). This was rather more than three months' salary, but he still had most of the £96 which his mother had given him to come out. Eventually, he decided that he would put down £25 of his own, and pay off the remainder, as a loan, over six months.

The motorbike arrived – it was a 500cc side-valve engine (rather bigger than it need have been!) and had acetylene lighting. It was Bill's pride and joy.

Not long after the motorbike arrived, a rather official-looking letter arrived from Kandy. Bill opened it with some curiosity – who could be writing such a letter to him? It turned out to be a summons for jury service at the Supreme Court in Kandy in four weeks' time. This seemed to be an extraordinary summons – he was only nineteen at the time and had only been in occupation of his bungalow for three months. His name must have come up for service as soon as it was registered. He rang Prior. 'Don't worry,' he said. 'This is nonsense; we'll get you an exemption.' However, the Court was adamant – Bill would have to serve. Secretly, he was rather intrigued by the idea. It would be a new experience.

On the due date, Bill told Banda to pack his grey flannel suit, and the rest of his kit, in a small suitcase, and put it on the ropeway for Dambatenne factory, where the bike was kept. Bill rode Eros up to Badger's Patch, and, after telling Raman to keep Eros exercised while he was away, walked down to the factory. The route to Kandy took him through Newara Eliya where he had been with Guy Simpson some months before, the scenery of which continued to thrill him. The jurors were billeted in Kandy according to their 'status' – Bill, being a European, was in the top category and was allocated accommodation in the Queens Hotel, with a room overlooking the lake. Luxury indeed!

Next morning, the court did not sit until 11 a.m. but he was required at 10.30 a.m. This still left him time to walk round Kandy and pass the entrance to the famous Temple of the Tooth.

Bill walked up the steps of the Supreme Court with a sense that combined both apprehension and anticipation – there was a tingling feeling running up the back of his legs. He wondered what it would be like were he to be a prisoner, rather than a juror! In the jurors' room he found to his surprise that he was the only European among them – all the others were either Indians, Sinhalese or Burgher (i.e. descendants of the Dutch and Portuguese, who had occupied Ceylon before the British. Most of these, though retaining their foreign names, had remarried into the local population and were now coloured. However a few, like the Barbers and Gratiens, had retained their pure ancestry).

Soon a court official came in. 'You are to be jurors in the court of Mr Justice De Silva,' he said, 'and you are to bow to him on his entry. Will you now please elect your foreman.' Without hesitation, they all elected Bill as foreman — quite a responsibility for one so young. Then they entered the Court, which was already filled — awaiting the arrival of the judge. Bill had no knowledge of court procedures, but he recognised counsel for the prosecution as being Noel Gratien, whom he had met when he played rugger for the Kelnie Valley club. Noel had played for the CR & FC which was the club for non-whites in Colombo. They had had a good evening afterwards — could he perceive a slight wink in Noel's eye as he sat down?

The judge arrived, clad in red and white, and sat stiffly as the court rose to bow. And so the trials began. Gratien opened the first case which concerned a Sinhalese woodman who was alleged to have murdered the husband of his lover — a common enough type of case. Evidence was produced by the Ceylon police, clad in khaki with Australian-type hats. The proceedings were in English with frequent interpretations into Sinhalese, which was an entirely new language to Bill. Gratien was brilliant and brutal, quite outshining Mr Ratnatanga who appeared for the defence. Bill made a mental note that this was not something that should be given too much weight in the verdict. The case lasted two days. Witnesses were produced for the dead man's relatives, but the accused seemed to be a 'loner' for whom no one could be found to speak. The law in Ceylon followed the Roman Dutch law, but, to Bill's inexperienced ear, there was not much difference from English law. The judge's summing up was fair and clear, gaining Bill's admiration. The jury took two hours to consider their verdict. Bill asked each member individually, for their views, and in the end there was a clear consensus for the verdict of 'guilty', to which the death sentence applied. Bill trembled a little as he rose to say, 'Guilty, my lord.' The judge looked sombre as he donned his black cap.

The jury service lasted a fortnight. During that time the court heard two more murder cases, one of which was acquitted, three cases of alleged rape, only one of which was proven and two of alleged armed robbery. Bill knew very little about the sexual relationships between men and women. The brutal nature of the rape cases were a revelation to him, leaving him ashamed of human nature. He gained the impression that sex for lust must be an entirely different sensation to that of mutual love. He hoped that, one day, he would find the 'right' person.

When Bill returned to Mousakellie he found that the Teamaker had been complaining about the quality of leaf that he was obtaining from the Division. Bill immediately asked Kadiraveil what this was all about, to which he replied that the condition of the leaf was no different from what was normally sent.

Bill mentioned this to Prior, and said that he thought that it might help if he could have a session at the factory with the Teamaker, to establish the cause of his complaint. Bill would also like to find out more about the process to teamaking. Prior was glad that Baker was showing this interest in the factory, and arranged for him to spend some time there.

The Teamaker was most helpful. As soon as the leaf arrives at the factory it is carried up to the withering lofts which form the main bulk of the factory. The leaf is then spread on hessian tats built horizontally from floor to roof of the loft with just enough room between the tats for the leaf to be spread or knocked down after it is withered. Once in place, warm air is blown by fans to dry the leaf.

The art of knowing when the leaf is withered is one of the crafts of the teamaker. The time taken to wither depends upon a number of factors such as the wetness of the leaf on arrival, the humidity of the atmosphere and even the age from pruning of the field from which it came. Usually, the leaf is withered at some time during the night. In any case, room has to be made for the next day's leaf. In very wet conditions the warm air may have to be increased. It is not often, therefore, that the teamaker gets a full night's sleep!

Once withered, the leaf is knocked onto the floor by long staves before being channelled to hoppers leading to the rollers. These are stationary machines with horizontal plates with radial grooves rotating in opposite directions, which cut and twist the leaf before passing it to the driers. The drying process is done by spreading the rolled leaf on trays which pass and repass slowly over heating elements until drying is complete and all moisture removed. The final stage is the grading of the tea, when it is blown over riddles. The main body of the leaf falls first as Pekoe, that with small particles including most of the tip then grades as Broken Orange Pekoe and finally the dust and smallest segments grades as Dust or Fannings. This latter, originally regarded as being inferior, is now popular as 'Quick Brew', but the speed of infusion seems to be at the cost of quality of the tea.

The coming of 1934 saw Bill becoming engrossed in his Division at Mousakellie. He liked to feel that he was now accepted by Kadiraveil and the other kanganies, and was proud of the appearance of the tea and its surroundings. He looked forward to Paddy Prior's visits, which gave him an opportunity to talk, and to hear what was going on in the world outside. He was sorry when, as Prior's reliance upon him grew, the visits became less frequent.

Bill's isolation was marked. He was two and a half miles and nearly an hours' journey from the Bungalows at Dambatenne, and two miles from Bandara Eliya. But there was always the telephone (which usually worked) and in any case, he was too busy to think much about being alone.

Monty Villiers heard that he wanted a dog. Someone at the Haldumulla club had had a litter of 'nearly pedigree' Cocker Spaniel puppies. Would Bill like one? Bill was delighted. In due course the post coolie arrived with a small furry head peeping out of a sack. Bill threw his arms around it, and it licked his face. He called it Kim, after Rudyard Kipling's tale (although it was a bitch!) It was to be Bill's constant companion, at first following him from room to room in the Bungalow, and then on his travels on the estate.

Bill's eye caught an advert in the *Ceylon Times* for a Philips Very High Frequency radio which was able to pick up broadcasts from England on direct transmission; a claim that seemed almost incredible. He ordered one, and, in due course, it arrived at the factory. Bill would not trust it to the ropeway, so the beef box coolie carried it up on his head. As he unpacked the set and rigged up the copper wires for the aerial and earth, he wondered whether he would really hear programmes from six thousand miles away. He was never to forget the first time he tried to tune in. There was a loud whistle which ascended in pitch as he tuned the waveband. Then, to his utter joy, a voice came over. 'This London calling, London calling, on nineteen metres VHF.' Good old England!

There were two batteries in glass containers, which lasted about two weeks. They were recharged at the factory.

Dambatenne was in the Uva Province of Ceylon, the county town of which was at Badulla. It was also the home ground of the rugger club. Sadly, with his new job, Bill was rather reluctant to commit himself to a full season's programme in his first year. Later he thought that Paddy would not have objected, but by then the teams had been made up and Uva had a scrum half. Nevertheless he joined the Haldumulla Club, where he played tennis. This was a game in which he had never been coached and he was not a 'natural'. His game improved gradually! The evenings were spent drinking, dancing and playing bridge. At first he was plagued by an initial shyness, but as time wore on he realised that this was to be the pattern of his social life, and the sooner that he threw himself into it the better. Someone organised a mobile treasure hunt with clues scattered around the countryside, which Bill on his motorbike won easily. He was given the next one to prepare – he had entered on the first rung of the social ladder!

Bill had now been on Mousakellie for about six months. He enjoyed his daily rounds with Eros, who was an ideal estate horse, willing and sure-footed. He felt confident on him even on the narrowest and steepest of rugged paths. It therefore came as a shock to him when one day Prior rang up to say that Eros was being transferred to another estate in the group. The horses really belonged to Mr Doudney the Managing Director, who owned some of them to race in Colombo. He had an Australian mare which needed resting; she was being sent up to Mousakellie as a replacement.

Bill wondered what this was really all about. Mousakellie was hardly a place for a full-blooded racehorse, nor was he sure about the reference to 'a rest'! He doubted whether Eros would have regarded his job in that light. But theirs not to reason why! Bill said goodbye to Eros with a frog in his throat; Raman showed obvious distress.

Soon Mary arrived. She was a young horse for estate work, five or six years old, a brown chestnut with bright, roving and rather nervous eyes. Somewhat ominously, she was being led on a snaffle. She had never been on estate roads before, and had not enjoyed her journey up from Colombo. Bill met Raman on the muster ground and they decided that they would take it in turns to lead her round the ground to try and get her accustomed. It was some time before they could get her to stand. Bill tried to mount. Immediately she lashed out with her foreleg, wheeling and snorting, just missing Bill in the process. Raman was obviously terrified. They tried again with the same result. Finally Bill said to Raman, 'That will do for today. Take her up to the stable and we will try again tomorrow.'

The next day Raman led Mary round the estate following Bill as he went. They had to take care as coolies passed on the way, warning them to go on the outside of the road against the fall. It was three days before Bill succeeded in getting on Mary's back. She immediately tried to bolt, causing Raman to let go her head. Bill had to pull hard on the snaffle to regain control – even so she was walking sideways and prancing alarmingly in a confined space.

Bill took to going out to the stable in the evenings after Raman had left. He talked quietly to her and offered her sugarlumps which, apprehensively, she accepted. Gradually the battle of wills eased. In order to mount, Bill had to climb from a convenient roadside rock, jumping onto Mary's back before she had time to respond. After a week, when Bill thought that things were getting more relaxed, Raman took Mary down to the bottom of the estate where Bill met them for the ride back. As usual Bill spoke to her, rubbing her nose and patting her before mounting from the rock. Before he had time to gain the stirrups Mary suddenly lifted both hind legs, bronco fashion, sending poor Bill flying in a somersault over her head. Having turned completely head over heels he landed, gracefully, feet-first on the hard stone road, but it was a nasty moment – there were no hard hats in those days. This incident was, perhaps, Mary's final capitulation. From then on, although still nervous, she became reasonably manageable.

One July evening, after he had completed the checkroll and Banda had cleared away the dinner, Bill was lying with his feet up on the long chair. He was listening to the BBC World Service on his new radio. The cricket season was in full swing. Sussex were batting against Surrey in the County Championship and had made 171 for 3. Owing to the difference in time, it was now dark outside and misty rain was falling.

Suddenly there was a knock on the verandah door. With some surprise at who this could be at this hour, Bill opened the door to reveal, on the threshold, Lydia Prior, dressed only in a blouse and khaki trousers. She was hatless, wet and bedraggled. 'May I come in?' she said.

Bill was too surprised to say very much. He showed her to a chair and put his bush jacket around her shoulders. She began to cry. Bill shouted Banda to bring some coffee. She was still sobbing when he entered to put the drink on the table beside her. As he left, Bill caught a look of disapproval in Banda's eye.

Bill left her to regain her composure before she explained this extraordinary appearance. It was unheard of for a white woman to be out on the estate, let alone the distant Mousakellie, by herself at night. And as for a bachelor's bungalow . . . well! There had been a Christmas party at the Big Bungalow to which he had been invited, and Prior had taken him through after a meeting at the office on a number of occasions, but he hardly knew Lydia.

Gradually her story emerged. Paddy had been married before and had a son, but the marriage had not been a success. When the wife took the boy Home to school she had not returned and they were divorced. Lydia was the daughter of a local rubber planter down in the low country. She was fifteen years younger than Paddy. Her father was something of a recluse and had never taken her to England. Indeed, after a visit on his first leave, he never went again. She had met Paddy while she was working as a secretary in Newara Eliya. They married after a whirlwind courtship. Since then, Paddy had never ceased to make her feel inferior and inadequate.

Up to this point, Bill had listened politely and respectfully. He liked Paddy as a man, and saw no reason to hear such disloyalty from his wife. He made for the telephone, but Lydia jumped up to head him off, throwing off the bush jacket and shaking her tousled hair. 'No, no!' she said. 'Don't call him!' She was becoming hysterical again, so Bill sat down. It seemed that Paddy was shortly going on leave and wanted to take her with him. He was boisterous and neglectful with his friends – she could not face being left to her own devices, and even laughed at, when they got to Ireland. She had refused to accompany him on leave. There had been a terrible row and she had walked out leaving her daughter, Monica, in bed.

Having got all this off her chest she became a little more rational. They talked quietly for a while – about the servants in the Big Bungalow, about her daughter, and, strangely, about Paddy's horse; the hysteria had passed – she was asleep.

Bill turned the handle of the telephone box, as slowly and quietly as he could, hoping and praying that there would be an answer. A voice replied, 'Yes Master?'

'Peria Dore, Peria Dore, *suruka, suraku*.'

An anxious voice came on.

'Baker here, sir. Mrs Prior is here.'

'Thank God!' said Prior. 'How is she?'

'She is asleep at the moment.'

'Will you see that on no account does she leave your bungalow? I will be up in under the hour. Perhaps you will send your boy down to Kadiraveil and ask him to get four good coolies to make a carrying-chair by strapping two poles to an armchair. They will then carry my wife back here. I will pay them two days' 'names'. They will need hurricane butties.'

'Right, sir,' said Bill. 'We will have the chair ready by the time you get up here. Your wife is thinly clad. May I suggest a raincoat and blanket? It is still raining up here.'

'Thank you,' said Prior.

As he was about to put the receiver down, Bill interjected, 'Are you sure, sir, that you would not prefer to stay the night up here? You could both have my bed . . .'

'No, thank you,' was the reply as Prior put the receiver down.

Bill wondered what effect this episode would have in his relations with Paddy Prior. The subject was never mentioned again, and if Prior was embarrassed, he did not show it. Bill surmised that the less said, and thought, the better.

Nearly all planters belonged either to the Ceylon Planters' Rifle Corps, (CPRC) or the Ceylon Mounted Rifles (CMR). Bill joined the CPRC and enjoyed the fortnight in camp at the hill station at Diyatalawa, not far from Haputele. There were also weekend camps in different parts of the island. A highlight of their training was when they joined the ships of the Royal Navy on landing exercises. There was plenty of opportunity for rifle shooting on the ranges. In the evenings they had singsongs led by RSM Hermes, the paid sergeant instructor, in which the officers joined them.

It so happened that Diyatalawa was also the hill station for the Royal Navy. As this was not far from Dambatenne, local members of the CPRC used to be asked to functions at the Station. Bill had several memorable parties there, and in return asked some of the midshipmen back to his bungalow for a taste of the planter's life. They used to marvel at the isolation in which Bill lived, but also envied his independence.

Paddy Prior had duly gone home to Ireland, taking Lydia and Monica with him. The acting Peria Dore was a Mr Nicholson from one of the low country estates. On the whole, not a very remarkable man and certainly not the character of Paddy. After an initial visit to Mousakellie, he left Bill much to his own devices. His attitude was that 'he was there if wanted'. A somewhat negative approach, which Bill took as a compliment, but he missed Paddy's more frequent assignations.

Although she had now settled down to the routine of estate work, Bill continued to be careful with Mary. He continued to talk to her in the stable in the evening, and she responded to his visits, and to the odd lump of sugar. He took something of a pride in his training of her. Nevertheless she could still be awkward at times and disliked, for example, the sudden appearance of a coolie around a corner of the road.

One morning, just before Christmas, Bill was riding Mary up towards Badger's Patch to see some coolies working there. The road zig-zagged from hairpin to hairpin as it climbed to the gap. Raman had taken a short cut up through the bushes to meet them at the top. The coolies had not seen Bill as he approached and were chattering away amongst themselves as they weeded beneath the bushes. Suddenly a coolie must have dislodged a small rock, which came hurtling down towards the road, then another, then another. As they fell, just in front of her nose, Mary's reaction was immediate. She raised her front legs in a frantic snort and whinny, nearly unseating her rider. Then, without warning, she walked backwards and outwards to avoid the danger. First her back legs were over the edge – then her whole body was falling.

With an instinct of self-preservation, Bill grabbed the stonework on the edge of the road; his hand held and his feet came out of the stirrups. He could just see Mary's anguished eyes as she rolled helplessly downwards, her reins and stirrups flying in all directions. Finally she disappeared over the cliff below. It had all happened so quickly that Bill could not believe it. He looked at the blue sky above him, the sun, the trees and the road beneath his feet – he was thankful to be alive. The coolies were gathering around him – Raman came tearing down through the bushes – '*Dore, Dore, ningula seria tan – seria tan?*' his anxious face peering up at his master. Bill looked at him with a half-dazed expression. 'We must go down and find her,' he said.

And they did. She was lying on her broken back, hooves projecting skywards, with her head strangely peaceful, lying in a ravine some one thousand feet below.

Bill would never forget those few minutes. There was nothing that he could have done, but he felt a great sense of loss – and even one of guilt. He went to the ropeway telephone; the Teamaker told Nicholson who came straight up.

'Get on my horse and ride it up and down past the point where she went over,' he said. 'Don't get off until I tell you to do so.' Hard words, but without them Bill would have found it hard to pass the point in future.

Looking back later on the event, Bill found a certain inevitability about it all. Whoever took the decision to send a young racehorse up to a young rider on a Division like Mousakellie must be held partly responsible. It was significant that although Mary came up 'for a rest', she was never recalled. The result might have been worse. The working party that Bill had appointed to recover the saddlery and dispose of the body had done its job; Mary had been buried close to where she fell in a bed of quicklime.

Bill was again without a horse. It was to Lipton's credit that he was sent, the next week, a grey Arab mare called 'All Clear'; twelve years old and much more amenable. Bill felt confident on her and Raman liked her. Perhaps her only vice was that she didn't like getting her feet wet and used to jump over the Irish drains rather than walk them!

Bill had been writing home regularly to Dolly. He looked forward to her letters and to hearing about his old friends. She sounded cheerful enough, but he didn't like to feel that she was alone there. Paddy Prior had returned from leave with Lydia and Monica. They all seemed to have survived his relatives in Ireland. One day, when Paddy was on his rounds, Bill, on sudden impulse, said, 'My mother is alone at home, sir. I have been wondering whether there would be any objection to her coming out to stay with me?'

'Why, of course there would be no objection!' said Paddy. 'The Bungalow is not marvellous, but if you think that she would like to come, by all means ask her.' Bill thought of saying that when Dolly married his father, they lived in a log cabin in the Rocky Mountains, but decided not to mention it – the circumstances were rather different!

Bill wrote off to Dolly at once. It is probable that Dolly had been a little alarmed by some of the accounts in his letters. She would be delighted to come out to Ceylon for a while. It would take her a little time to pack everything, and she would sell Upper Cottage. She hoped to be able to come early the next year (1935).

Meanwhile, Bill had joined the Uva Rugger Club. This was one of the principal provincial up-country clubs with a fine pitch and clubhouse at Badulla. The rugger season was one of the social events and was supported by many of the planters and their families. Bill played in the trials, but the previous season's scrum half, John Corlett, was still available. Sadly, Bill didn't get a place. He went down to Badulla on his motorbike to watch Uva play his old friends from the Kelnie Valley. Keeping a critical eye on Corlett, Bill thought that he was too slow, and not particularly fit.

Bill heard nothing from the Club for six weeks. Then, with only two matches left to play, the post coolie brought a card. 'You have been selected to play for Uva against Dimbula on 20 November. Please confirm.' Bill was over the moon. He had kept himself fit – in case. It turned out that Corlett was sick, but Bill, having got his place, didn't intend to lose it! Uva beat Dimbula quite easily, and he was selected again for the final match against the CH and FC (Colombo Hockey and Football Club) in Colombo on 14 December. The town wallahs put up a strong side and won 21–18, but Bill scored a try for Uva and was satisfied with his performance. Bearing in mind the length of his journey back to Dambatenne and his walk afterwards, he left early.

Bill received a cable. 'Arriving Colombo 15 February. Love, Mater.' She was fifty-eight at the time.

He had an immediate problem about transport. He could not contemplate taking her from Colombo all the way up-country on the back of his motorbike! He set off for Colombo the day before she was due to arrive. After some haggling, he was lucky to exchange his motorcycle for a two-seater Austin Seven with a 'bubble'-type boot. It was a somewhat ancient vehicle, dating from about 1926, but it was a good runner. It was his first car and had cost an additional £40 over the value of the bike.

Bill couldn't help comparing the emotional meeting with Dolly on the jetty with the restrained and formal reception that he had received from Leicester some twenty months earlier. Dolly was looking brown and well after her sea voyage. He was thrilled to see her looking so happy. The purchase of the car had been completed with the minimum of formality – just a word with the National Bank of Colombo. Soon they were bowling along on their journey up to Haputele. There was no synchromesh on the gears, so he had to double-declutch. He also found that she would run with the ignition fully advanced on the flat low country roads, but as soon as they began to climb he had to retard by easing back the lever on the steering wheel.

They had so much to talk about on the way up, and Dolly was so enthusiastic over the scenery, that the time on the journey passed all too quickly. Bill had asked Kadiraveil to arrange for four strong coolies to meet them at the factory with a carrying chair, for which he would pay privately. Dolly's *saman* would go up on the ropeway to the muster ground, and from thence to the bungalow on the head of the garden coolie.

And so the cavalcade set off. Dolly perched, somewhat precariously, in the chair on the shoulders of the coolies, followed by Bill on All Clear with Raman bringing up the rear. They stopped, as usual, at Badger's Patch (Bill was a little reticent about his experiences there. How different it all was on this beautiful sunny day!) Banda was standing outside the bungalow to meet them, looking smart in his white jacket and *veti*, his somewhat staid expression relaxing. 'Welcome, mistress!' he said. Dolly showed little formality in greeting him. There were some jokes – some of which were at Bill's expense! Banda was glad to have a lady in the house: it gave him a certain status in the boys' hierarchy.

During his next visit to the Division, Paddy Prior called in to pay his respects. Banda brought them all coffee; it was quite a social event. Paddy found Dolly more 'worldly' than he had expected; there was no lack of conversation. Paddy was prepared to have to make apologies for the Bungalow, but Dolly had the knack of talking about the good points: the wonderful situation looking out over the mountains, of being looked after by Banda, and a mention of how enthusiastic Bill had been in his letters, all of which went down well with Prior.

They had given Banda the job of finding a houseboy. After a week he produced a young Tamil by the name of Velu who was in his mid-twenties and had quite good references. He was engaged without further ceremony.

He was to be paid twenty rupees a month and the cost of half a bushel of rice. Velu's English was not very good and he was rather 'green' at first, but the main thing was that he got on with Banda.

There was now quite a little community around the Mousakellie Bungalow. In addition to Banda, Velu and Raman the horse-keeper, there was Pichi the garden coolie, who, in addition to running the small garden, used to carry out all the odd jobs like finding people when they were needed or fetching things from the factory. He was an expert at carrying heavy objects on his head! Outside the veranda windows there was a large coffee bush, a relic from when the whole estate had grown coffee with such success in the last century. A disastrous blight had killed off the coffee within a matter of years, ruining many of the old planters, but with typical fortitude the land had been replanted, with, perhaps, the even more valuable tea. The coffee bush had grown into a small tree and after the monsoon was covered in berries, which at that stage were bright red. Banda used to gather them. Some he stored for future use, taking them out as required for roasting; a process that gave rise to a wonderful aroma.

Even in the nineteen-thirties, Ceylon was not without its political troubles. There was a continuing friction between Tamils and Sinhalese, upon which some hard-line agitators used to build. The police, who were mostly Sinhalese, did their best to subdue these factions, but on occasions the white population had to lend a hand. Such an incident arose not long after Dolly's arrival. The local planters were called out 'in aid of the Civil Power'. Harry Carter, the SD who had replaced Monty Villiers on the Dambatenne Division, Bill, and Philip Maudsley, an SD on Monrakandy, were to take their CPRC rifles, forty rounds of ammunition and their horses to Haputele, reporting to the Police Superintendent. This provided quite a little excitement as they rode through the villages on the way. On reaching Haputele there was a contingent of about thirty planters who took it in turns to patrol, in groups of six, the troubled area. After two days the police had regained control. The planters returned to the estate. All was well. But it served to remind the Europeans that they were sitting in judgement upon a vastly predominant native population and, to some extent, imposing a civilization that was foreign to the locals.

Bill had not been very happy about leaving Dolly up on Mousakellie by herself, but he had no choice. However, he was glad to be back.

Bill was now an official Bugler in the CPRC. He used to play The Last Post from his bungalow as he went to bed at night. He never knew quite why. Perhaps, as the notes reverberated around the valley, it gave him a feeling of English individuality. He never asked what the coolies felt about it – and they, politely, never told him!

The rugger season came round once more. This time Bill was the undisputed scrum half and there was a full fixture list, meaning that Bill was playing every other weekend. He had to have a good jab of tetanus vaccine before the season started because sheep had been grazing the pitch beforehand! Paddy Prior continued to be very good about letting Bill off – there was no doubt that Liptons received a certain amount of kudos through his being in the Uva team, but once or twice he sensed a certain jealousy on the part of the other SDs at the amount of his leave. Sometimes Dolly came with him, in which case she used to walk to the factory and have a carrying-chair on the way back. Bill always left the club early after the match when she was with him. One such event was against the Dimbula team, when Leicester and Diana Green were watching the match. It was the first time that they had met Dolly for many years – it was all pleasant enough. Bill was glad that it was so. Whatever he might feel himself, the past was over and Bill was now established in his career.

The work on the Mousakellie Division was going well. The yields of leaf had been increasing since Bill came, which made both the coolies and the Company pleased. Bill was insistent that the major works, such as draining, weeding and pruning, were all done effectively and to time. He also kept a careful check on the fertilizer, to see that it was all applied and properly distributed. This was not always an easy task; it required spot-checking of the bags in the fields and of the level of application.

The result of this was more leaf which had, rigorously, to be plucked to time. If extra time was needed then it was paid for by the additional leaf credited to each plucker. He found that Kadiraveil, for all his dourness, was glad of this general improvement in production, though if he thought that the coolies were being pressed too hard, he would say so.

Bill always found the morning muster a good indication of the well-being of the coolies. If work was unpopular, or if the coolies were being pressed too hard, then attendance would fall, causing a need for discussion with Kadiraveil and the other kanganies. After muster, Bill would take the reins from Raman and ride All Clear back to the bungalow in the fresh morning air. This was a favourite time of day. All Clear liked it too; he would trot eagerly home towards his stable. As he neared the last section of the road he would break into a gallop, literally slithering his hooves over the stonework in the Irish drains. There was a final bend which he took at full speed, with Bill leaning almost at right angles to keep his balance. This became quite a morning spectacle which some of the children lined the path to watch! Bill did nothing to encourage the horse – it seemed that he did it out of pure *joie de vivre!*

George Villiers, the SD on the Bandara Eliya Division, had been transferred to a more senior post on another estate. As a replacement, Liptons had engaged a Cornishman by the name of Trevellyan, who was on probation. Bill never discovered his background or what training he had had; at all events, he was not a success. He had an inflated opinion of himself which he tried to impart both to Harry Carter and to Bill. It seemed that he had very little knowledge of Tamil or of the Tamils' customs and he soon got across the Head Kangany. Finally Prior, in a fit of frustration, told him to pack his bags and three months' pay.

That evening, Paddy Prior rang. 'I want you to go to Bandara Eliya.'

Bill was not sure whether to be glad or sorry. It was a promotion: Bandara Eliya was a larger Division with eight hundred and fifty coolies and had some of the best and most valuable tea in the world. It was planted on a plateau at some six thousand feet with an ideal climate for growing tea. But sadly, because of its flat topography there was no horse! However, the Bungalow was slightly better (but not much) than Mousakellie and the access, though still about one and a half miles from the nearest cart road at the factory, would be easier, especially for Dolly. Anyway, of course, he had to go.

So it was goodbye to Mousakellie. There had been something romantic in its remoteness and its challenge which would never quite be achieved in his future billets. Somehow, Mousakellie was personal to Bill – he didn't like to feel that someone else would be in the bungalow and in charge of the Division. He said some kind words to Kadiraveil and the kanganies, who did not, however, on this occasion, attempt any kind of festivity! There was a small gift for Raman, the horse-keeper, and for Pichi, the garden coolie. Bill sought out Muttusami the road-mender for a yarn before he left.

Banda and Velu packed up their belongings which were carried up the back road between the two Divisions. With a slight attitude of defiance, Bill decided that he would arrive at Bandara Eliya on All Clear, even if the horse had to return to Mousakellie afterwards! Dolly came up by carrying-chair.

Bill could not fail to be impressed by the vast and level expanse of beautifully cultivated tea before him. No wonder Liptons, and Mr Doudney in particular, were proud of this showground of the planter's art. It went

Káliamma: ready for the day.

through Bill's mind that there was little opportunity to impress one's own individuality upon the place – no wonder the luckless Trevellyan had succumbed. The Head Kangany was Supramannian, of higher caste and appearance than Kadiraveil. He was both confident and diffident; Bill guessed that he would have to rely upon his judgement for a time. The Division carried a conductor: an educated Muslim who was in charge of the day-to-day work on the Division, and whose name was Mr Chad. Bill made a mental note that he would have to watch the division of responsibility between Supramannian and Chad. The Conductor was also normally responsible for keeping the checkroll – a considerable task in view of the number of coolies. On the other hand, Bill relied upon his keeping of the checkroll himself for the learning of the coolies' names and their gangs. For the first few days he would act as an observer, to see how things worked.

The bungalow was much the same as Mousakellie, except that there was piped water fed from a reservoir which was itself fed by water pumped by a Pelton lower down on the estate. There was still no electricity, but the bungalow was partly furnished and had a number of Aladdin lamps and hurricane butties. By the time Bill came in from muster in the evening, the two boys had things reasonably shipshape and Dolly had arrived in the carrying-chair.

Dolly thought that it would be nice to have a second dog as company for Kim; besides, she would like to walk round the estate while Bill was out, and it would be company for her. There was a lady in Bandarawela, not far from Haputele, who bred Cocker Spaniels from the 'Of Ware' strain. She had a litter just ready for sale, so Bill and Dolly went over to see them. The choice was difficult but one of the puppies cuddled up to Dolly, so, of course, it had to be that one. They brought it home and christened it Boots of Bandara Eliya which they registered as its kennel name. It became known as Bunty – Kim was uncertain of it at first, but soon adopted it as her own.

A new SD had arrived at Mousakellie, a young Irishman named Patrick Moore who, Bill gathered from the bush telegraph, was a distant relative of Paddy Prior. Pat was a nice lad with not much experience of tea planting but was eager to learn. He didn't play rugger! Bill used to invite him to supper at the Bandara Eliya Bungalow sometimes. There was another SD on an adjoining estate to the Division, named Ian McKenzie. The three of them would put the world to rights until late – Dolly would join them at first, before excusing herself and retiring to bed. Ian could, by some skilful and determined riding, get his motorbike to Bill's Bungalow – poor Pat had to walk!

It was one of the Lipton edicts that the Visiting Agent, Mr H. L. Doudney, should visit each estate in the Group at least once a year. Prior notice was given and there was a noticeable 'grooming' of the estate beforehand by the Superintendent. Although he was not exactly jittery, Prior wanted to know the programme for the day, where the coolies on each Division would be, and worked out a plan of where to meet each SD on the way round. This tour followed, no doubt, a fairly full session in the office to examine the accounts.

Doudney and Prior were due to meet Baker at the head of the ropeway on Bandara Eliya, from whence he would conduct them around the Division. They duly arrived and Bill showed them, with some pride, the major plucking fields. Both the Conductor and Supramannian, the Head Kangany were there and received suitable approbation for their work. Etiquette required that, when the SD was present, there should be no displeasure expressed direct to the kanganies. If such was needed, there would be some direct talking to the SD afterwards! In this case, there was none. This little custom was to stand Bill in good stead, later on, when he was in the Indian Army, where the same rule applied. It was a custom which some of the British officers failed to understand, and which sometimes led to quite serious incidents.

As the conducted tour continued, they came to the pruning field, where the core of the labour force were working. Doudney announced that he would like to go into the field to see the pruners and to talk to them. (He was quite good with his Tamil language.) Prior and the VA both got off their horses, handing them to the horse-keepers, to walk through the pruned tea until they came to the line of men working. As Doudney stooped down to speak to one of the pruners, the next man shouted at the top of his voice, '*Aiyo, Aiyo, Perisi Ten-i, Perisi Tenii!*' The man had put his knife into a nest of hornets!

Chaos reigned.

Bill shouted to the men to take off their *kumblies* (the blankets worn round their loins to protect them from the sharply pruned branches), lie down, and cover themselves. He pushed Doudney into a nearby ditch, grabbed a coolie's *kumbly*, and, with the three of them – coolie, Doudney and Bill – all in the ditch together, spread the *kumbly* over them. Just in time: the angry hornets were all over them. Prior was in an adjoining ditch, spread-eagled beneath the Head Kangany. The hornets vented their wrath upon the poor unfortunate horses, as hornets are prone to do. Doudney's horse (which was that belonging to the Dambatenne Division) was stung by two hornets in quick succession. With a loud whinny it reared up, releasing itself from the horse-keeper, and bolted back down the hill to Dambatenne.

Doudney was visibly shaken by the whole affair. When the hornets had departed, they assembled on the pathway. 'Thank you,' said Doudney, 'that could have been a nasty incident.' The hornets in Ceylon, though not very common, are particularly vicious. Four or five stings on one person can be fatal. The horse-keeper was despatched to find the horse, while the Visiting Agent's party ended up in the Sinna Dore's Bungalow to be entertained by Dolly while Banda was preparing tea. The incident caused much merriment later, but, at the time it was not very funny. Afterwards, it took a little time to restore the formality of the occasion.

The troubles between the Sinhalese and the Tamils were still rumbling in the villages, particularly down in the low country. Normally this was a matter for the police, but the Ceylon Planters' Rifles continued to be called out on occasions. They were 'put on show' in the area of the trouble, and certainly seemed to quieten things down. They then went back to the estates and, apart from some censored newspaper reports, tended to forget all about it. However, one night a stark reminder fell onto Bill's lap. Bill had gone to bed, as usual, after playing The Last Post on his bugle, and was sleeping the sleep of the just. Suddenly, there was a hammering on the front door which Banda failed to answer. Sleepily, Bill opened the door. An agitated Head Kangany was standing on the threshold. He was normally neatly and cleanly dressed; Bill was surprised to see him clad only in a loincloth and vest.

'Master, master, come quickly; a coolie has been killed in the lines.'

Bill scrambled into his shorts and bush shirt. Grabbing a hurricane buttie, he followed Supramannian into the lines, where there was commotion and wailing, with the women tearing their clothes. On a path just outside the lines lay a headless corpse – the head being further down the path in the ditch. Bill immediately ordered the kanganies to get the coolies back into their rooms and away from the body, which he ordered to be enclosed by a rope fence. He also told the kanganies that no one should leave the lines until the police arrived. Taking the Head Kangany with him, he returned to the bungalow and rang the police. It transpired that the dead man was something of a Tamil nationalist who had been down in the low country to whip up a counter to Sinhalese aggression against his kinsmen. It seemed that a gang of Sinhalese activists had followed him back to the estate, dragged him out and murdered him.

By now, the first grey light of dawn was appearing. Bill rang Prior. He was sorry to disturb him, but, perhaps he ought to know . . .

There was much police activity on the estate for the next week or so. A police sergeant even came to see Bill, but he never heard that the murderers were caught – the police were mainly drawn from the Sinhalese.

The Austin Seven 'Bubble' had been a great joy. In addition to his trips to Baddula, and the other rugger grounds, Bill had taken Dolly on a memorable run southwards through Welawaya and the jungle road to Hambantota, where they stayed in the rest-house. Suitably covered in suntan oil, they stayed on the beach and Bill had a swim in the rolling breakers. Fishermen went out in their catamarans made out of hollowed-out trees, bringing in loads of fish, much of which was sold locally for a few annas a pound. The rest-house keeper made the most marvellous fish dishes for which he was renowned all over the island.

Dolly had taught Bill the rudiments of Bridge – she was a good player herself, and Bill's father's brother Archie used to play in championships. Bill was keen to try himself at the Haldumulla club where he and Dolly played with people most likely to be tolerant! At least he had started. His tennis also began to improve, but he would never make his name in that sport.

With the rugger season approaching once more, Bill decided that for all its individualism and stout-heartedness, the Bubble was not quite up to the demands being placed upon it. The agents in Colombo were advertising new Austin Seven open tourers for 1,430 rupees (£110). By now he had enough in the bank. The firm offered him a trade-in on the Bubble of £35, so he was the proud owner of a *new* car! Dolly was delighted. Though she did not say so, she had always found the Bubble rather hard on the springs!

Pat Moore was always grateful for Bill's help and they became quite friendly. Together with Ian Mckenzie, they were often in each other's bungalows. One evening Pat announced that his younger sister, Sue, was coming out from Ireland to live with him. Some girl, thought Bill, who wondered whether she had been warned about the remoteness of Mousakellie.

In due course Sue arrived. She joined in with the generally masculine activities both on the estate and at the Haldumulla Club. A rather thin, almost skinny, lass with a high-pitched voice which concealed a determined and purposeful nature, she was good at tennis and was generally popular with everyone. To Bill, she soon 'settled into the scenery'. The rugger season had started; he had other things on his mind.

Bill missed not having a horse, but the road up to Bandara Eliya, although rising by some one thousand five hundred feet, had an easier gradient than that of the access to Mousakellie. Bill thought that, by improving one or two bad patches, he would be able to get a motorcycle up to is bungalow. He discussed this with Prior, who agreed that the stonemason should devote a week's work. Bill saw an 3½ h.p. BSA side valve advertised in the paper for 250 rupees. He bought it – a beautifully simple machine that he could easily maintain himself. Soon he had it roaring over the Irish drains and bumps – what a difference it made!

As the rugger season progressed, Patrick brought Sue down to the Badulla Club where there were, in addition to the two teams, a number of young SDs and their wives or sisters. Sue joined Bill's party and they danced together. Now that Bill had his motorbike to get back to Bandara Eliya from the factory he was sorry for, and rather admired, Sue for being prepared to walk home, albeit with Pat, in the small hours.

Now that their bungalow was closer to the cart road at the factory, Dolly found that she could walk down and back without the need of a carrying-chair, provided this was not done in the heat of the day. There was a little Anglican church in Haputele, which had services once a month taken by visiting padres from Badulla, Newara Eliya, and, occasionally from as far away as Kandy. They usually stayed to have a word with the padre afterwards, and sometimes went with him and other members of the congregation to a neighbouring bungalow. Sometimes they were joined by Alastair Gorman, who was now the Senior Assistant at Dambatenne.

On a rather remote part of the Bandara Eliya Division, on land slightly more elevated than its surroundings, there were some old plumbago mines. These had been worked at the end of the last century, to provide black lead for pencils and for mixing with clay for making crucibles, leadware and so on. When the material had been worked out, the caves, which mainly ran along the flat, into the hillside, had been left open. They had now become colonised by thousands of bats, which in the evening could be seen in swarms over the caves. Inside, the roofs were alive with the creatures. Their long hairless legs and necks and beady eyes gave Bill the creeps. On the ground, the floor of the caves was covered with evil-smelling guano, which was supposed to be a good fertilizer. No one had suggested that this should be gathered for spreading on the tea. Bill certainly wasn't going to suggest it!

However, Alastair's mother and seventeen-year-old sister, Diane, had arrived in Ceylon for a few weeks visit and were staying with Alastair. The young Diane rather fancied herself as a naturalist. On hearing about the plumbago caves, she insisted upon exploring them. Bill, somewhat reluctantly, was deputed to accompany her. Diane and the mother duly arrived, and while the elder women sat down to tea, Bill set off for the caves. Diane was a reasonably pretty girl, but she had girded herself in a *kumbly* that she had obtained from the factory and had a large tea-towel affair over her head. To Bill's dismay, she obviously meant business. They both had hurricane butties which they lit on arrival at the entrance.

'How far have you been in?' asked Diane.

'Not very far, about twenty or thirty yards,' said Bill. 'The bats are not very attractive creatures, and the smell is awful.'

'Oh, I'd love to see what it is really like inside,' she said as she entered.

As they got out of the light, the roof of the cave became covered with animals. Hanging upside down, they swayed from side to side, their small piercing eyes following the intruders. The females had their young clinging to them, issuing a faint 'yeeping' sound. The guano beneath their feet yielded to their tread; they were soon wading in the stuff. The passage turned at a sharp right angle excluding all light except that which flickered from the hurricane lamps. It seemed that the bats became bolder in the darkness and were now flying in the cave above their heads; it was only the hurricanes that kept them from flying almost into them.

'Are you sure that you have not had enough?' said Bill hopefully. His mosquito boots were covered in guano and every so often he felt excrement falling on his topee.

'No,' said Diane, 'I want to see how far the bats live in here.'

As she said this, a large bat that had been particularly restless swooped down on her, becoming entangled in her headcloth. At the same moment, as she looked up at the bat, she stumbled over a rock, letting go of the lamp as she fell. With a sickening crash the lamp fell onto the rock, spilling the paraffin which ignited around them. The effect upon the bats was immediate. Leaving their perches, they flew frantically around the passage, sometimes through the flames. Soon the passage was alive with bats flying in all directions. Bill put his arm around the girl as they crouched against the wall of the passage. He clung tightly to the remaining lamp, waving it at the animals if they came too close. The girl was shaking, but did not scream.

Gradually the bats returned the their perches. The flames from the lamp had died down. 'We will go back now,' said Bill. This time Diane did not demur.

When they arrived at the mouth of the cave, they looked at each other and burst out laughing – they were covered from head to foot in guano, even on their faces. They must have fallen into it during the melee.

Back at the bungalow, Diane's mother gasped. 'What have you two been doing?' she said, eyeing Bill a little reproachfully. They did not repeat the story of the lamp.

'The cave was full of guano everywhere,' said Diane. 'We couldn't avoid it.'

Dolly offered the girl a bath which she declined, saying sensibly that she had no clean clothes to change into. As they left, Diane's mother said, 'Where's the lamp you brought with you, darling?' There was a non-committal reply. As he lay in the bath, Bill wondered what would have happened if he had also tripped over the rock, and lost the second buttie – but he dismissed the notion. Enough that he would have to buy a new topee.

On one occasion, Bill and Dolly stayed the night with the Ashworths at Nanoya before continuing up to the northern part of Ceylon to visit the ancient rock fortress of Sigiriya, on the road to Trincomalee. This place held a fascination for Bill. They were travelling through the flat lowland jungle totally without any landscape features, when suddenly they came upon a huge rock with nearly vertical sides up to two hundred feet or more. The top was quite flat, giving a view from two to three acres in extent. The rock formation is a derivative of volcanic magma. It is said that the Sigiriya rock is the 'plug' of an ancient volcano, the sides of which have eroded leaving the plug resting at ground level. As they approached they called in at the rest-house to order lunch. The rest-house keeper was a well-known character who had an island-wide reputation for Indian cuisine. They would be back in about an hour.

A path led from the car park up to the rock. Then there was a climb up a narrow ledge cut out of the vertical face of the rock, with iron railings, none too securely fastened, along the lower side. Hairpins as the path reversed were formed on iron platforms jutting out from the rock. As they climbed, there was a magnificent view looking out over the tops of trees in the jungle, but looking down, one had a queer tingling sense in the soles of one's feet.

Then they came upon the famous Sigiriya Caves. Here the side walls of the huge rock were overhung, creating a dry open cave protected from the bright tropical sunlight. In prehistoric times there had been many ancient civilizations in Ceylon, of which the city of Anuradhapura is probably the best known. Here, even earlier, the cave dwellers had plastered the walls with lime and dung before painting some exquisite figures of hunters, animals and scenes in the jungle. There were also most delicately drawn figures of women in costumes of the time. Some of the original colouring, applied thousands of years before, still survived in the dry atmosphere.

When they had seen the caves Dolly said that she had climbed enough, so she would wait there while Bill went on to the top. After he had negotiated two more hairpins, he found that the path petered out. The rest of the way was by an iron ladder hanging from the side of the cliff! Undeterred, Bill climbed upwards, taking care not to look down. From the top of the ladder he stepped out onto a flat, unfenced and treeless plain surrounded by the jungle some two hundred feet below. It was a strange but exhilarating experience. Dolly was waiting patiently for him back at the caves.

They were late for lunch, but the rest-house keeper was expecting that they would be. They had the most wonderful curry, the fish caught locally and most of the other ingredients being fruits of the jungle. For sweet they had his famous rum omelette – four eggs per person done on one side only with the centre piled high with a mixture of raspberry jam and rum by the tablespoon. A well-earned meal long to remember!

It was dark when they reached their beds at the bungalow at Bandara Eliya. Bill was soon fast asleep. Dolly lay awake in bed wondering about the future. Clearly this kind of life could not go on much longer.

The rest of the year passed all too quickly. Bill had his best rugger season ever with the Uva Club. They won the Up-Country Tournament beating Kandy, Dickoya and Uda Pusselawa. Bill and Sandy Richardson were the two halves playing in the Up-Country trials for the All Ceylon final. Things were going well on the estate too. He got on well with Paddy Prior who was tolerant of his growing confidence and enthusiasm.

Towards the end of the year, Prior spoke to Bill on his round. There were to be some moves afoot. The second SD on the Pooprassie Group was leaving the company. Bill was being posted to Pooprassie in the New Year. Prior was sorry that he was going, but it would be to his advantage to get more experience. Shortly afterwards he received a formal letter posting him. He was to report to Mr M. P. Brazier on 4 January 1938.

Bill had been half expecting this. It was not a good thing for a SD to be 'brought up' on a single estate, or mainly under the supervision of one Peria

Dore. However, it was well known that there was no love lost between Prior and Brazier, and, by the same token, an SD who was liked by one was unlikely to be popular with the other.

Bill rang Brazier immediately after Christmas, introducing himself. He mentioned that he had his mother living with him, saying that before arriving on the 4th, he would like to come over to see the Bungalow. Brazier was not enthusiastic, but did not demur. Having told Prior that he was going, he took Banda over in the car leaving Velu with Dolly. The Bungalow, called simply, the Middle Division, was quite a pretty little place, fairly near the cart road, which was a great advantage, and entered by an avenue of tall eucalyptus trees. They climbed up the steps to look around inside. The furniture was simple, like that of the other SD Bungalows.

Bill went on up to the Big Bungalow to meet Brazier, who was a dour Scot. In his late fifties, he was one of the old school of planters who had seen hard times. He was now awaiting his retirement. Mrs Brazier was a pleasant woman, well-dressed, talkative, and rather apologetic of her husband. Bill did not take to her, though he could not quite say why.

There was not the air of bustle of efficiency about Pooprassie that Bill had been used to. The Lower (factory) Division was in charge of John Digby, a bachelor in his early forties, whom Bill suspected had a Sinhalese woman 'looking after' him. Anyway, he didn't entertain in his Bungalow and wasn't very communicative. Bill decided that, so far as the estate was concerned, he would have to 'paddle his own canoe'. And there was much to be done. Bill soon found himself engrossed in learning all he could about his new Division.

One particular feature of the Middle Division was to Bill's sheer delight. On the eastern boundary of the estate there was a wall of limestone cliffs with bougainvillaea, lantana and other tropical plants hanging from its face – it was a natural garden of its own. Even more intriguing was a little-used stone stairway leading up through the wall to open prairie above. Running down through this was a crystal-clear stream which cascaded into a limestone pool before spreading out over some rocks worn smooth by the running water. It was a perfect place to lose oneself for a swim in the evenings, and to lie on the rocks with the water shooting over one. Bill would spend many a weekend there, and sometimes Dolly would join him for a picnic.

Dolly liked the Middle Division Bungalow. Although neglected, there was a nice little garden. The *tota karen* responded to Dolly's interest. The eucalyptus trees cast a welcome shadow which enabled her to sit out during the evenings; she said that the aromatic leaves kept the mosquitoes away. One rather strange thing happened not long after they arrived. Both Dolly and Banda were meticulous about putting the living-room furniture straight last thing at night. One morning, when Banda brought Bill's early morning tea, he seemed a little upset.

'What's the matter, Banda?' asked Bill.

'Did master go into the living-room during the middle of the night, please? Well, table and two chairs have been put into the middle of the room. Master, please come and see.' True enough, there was a small side table that they had certainly not used the night before, together with two chairs, in the centre of the room.

'Well, perhaps you will put them back where they belong,' said Bill, in a matter-of-fact tone of voice. One of them must have left them there. He couldn't quite understand why Banda had been so upset.

The local Club was in a disused tea factory and had been dubbed the Delta Club. It was a jolly affair with a number of young people. In addition to the usual tennis and bridge, the young ones organised some unusual games to soak up their exuberance. One of these consisted of two teams of five, one at each end of the bare factory floor. To start, a cushion was thrown into the centre; the aim was to get the cushion to touch the far end wall on the opponents' side. However, it had to be held by the attackers at the time! The game was hilarious; not all that dissimilar from rugger. Bill revelled in the physical effort required and was much in demand. Strangely, no one seemed to get hurt; young bones are hard to break!

There was a family called O'Riorden. Both the husband and wife were what in England would be called 'County class', besides which they were a charming couple. Sion O'Riorden tried to coach Bill with his tennis: Bill tried, with not much success, to emulate the way Sion 'stroked' the ball, rather than hit it. There was a little improvement but much! It was not his game.

One Sunday there was an American-style tennis tournament at the Delta Club, in which the men players drew their lady partners by ballot. Once drawn, the couple played one set against each other couple. Sometimes, this arrangement produced some unusual results! On this occasion, Bill was playing with a Miss Sarah Jordan, a strong and rather belligerent lass who was the daughter of a planter in Gampola. Sarah rather fancied herself at tennis and was not particularly pleased at having drawn Bill as her partner, who, for all of O'Riorden's coaching, still needed to improve his game.

Owing to the number of sets to be played during the afternoon, they had to be on court early while the sun was still high and the heat oppressive. However, unlike some of the others, Bill was now used to being out in the midday and took advantage of the situation. They won the first three sets on the trot and Sarah began to applaud her partner, even though it was she who did most of the 'killing'. In the next set, they were drawn, a little unfairly, against a husband and wife who had drawn each other. Although their play was not by any means exceptional, this couple won by co-operation and anticipation of each other's movements to which Sarah and Bill had no answer.

However, by now Sarah was playing quite superbly; they took the next two sets to bring them to the final. The deciding game was against O'Riorden and a not very strong lady partner. Understandably, Sarah and Bill aimed everything that they could against her, but O'Riorden was irrepressible. At the net he took every return from his partner's serve and, with a mighty top spin, broke through on his own serve. The result was inevitable. Bill and Sarah were the runners-up.

'Well done, Bill,' said Sarah as they left the court, giving him a peck on the cheek. 'I misjudged you.'

'Thank you,' said Bill. 'You were brilliant, but that man O'Riorden is impossible! Now, how about a large lime and soda?'

In the evening the men were too exhausted to play the cushion game, so they all sat down to bridge. One young couple who were members had brought their five-year-old son, John. Jack Hooper was a keen bridge player, while his wife Maureen, either by inclination or because of her family ties, preferred to stay in the bar chatting. As the evening progressed, Bill was sitting at one table to partner the Club President's wife, Mary Loudon Shand, while their opponents were an Irishman, Barney Prescott, and his wife.

Between rubbers Mary asked Bill how he was getting on at Pooprassie. On hearing that he was on the Middle Division, Barney pricked up his ears. 'Oh no!' he said. 'That's the Bungalow that is haunted, isn't it?'

'What do you mean?' asked Bill, taken aback.

'Well,' said Barney, 'there was quite a to-do with you predecessor. The furniture kept moving about, or something.'

'Oh?' said Bill, deciding to keep his peace. 'I hope that does not happen to us!'

Bill's table was halfway through the second rubber when the door from the bar opened for young John Hooper, Jack's son, to enter. After looking around the room, he came straight up to Bill. 'Come on, Daddy,' he said. 'Mummy wants to go home.'

A roar of laughter went round the room. Jack and Bill had the same fair hair and blue eyes. 'We never knew about that,' said Barney. 'You *are* a dark horse, Bill.'

'Don't be ridiculous, Barney,' said Bill, but not without a slight blush, as Jack Hooper grabbed his son's arm. However, from thence, Bill became young John's 'second Daddy'.

On his way back to Pooprassie in the car, Bill pondered Barney's remark about the haunted Bungalow. The mere fact that the table and chairs had moved did not worry him; what did matter was that Banda seemed to be so distressed. He decided to have another word with Banda in the morning. He was asleep as soon as his head touched the pillow.

Banda woke him as usual for muster, but was in a state of great distress. The table and chairs had been moved again! 'Right, Banda, leave them where they are. We will talk about this when I get back,' he said as he dashed off to the muster ground.

And so they did. It transpired that the table and the chairs were not the only cause for Banda's distress. Other items in the kitchen had unaccountably gone missing, or been displaced. Even some of the master's washing had been found in an outhouse. However, what really distressed both Banda and Velu was that, since their arrival, they had been treated as outsiders; Banda had even thought of leaving his master if matters did not improve.

As Banda was talking, it became clear to Bill that all this was some sort of storm in a teacup that would have to be resolved, but why, and how? Giving Banda an assurance that he would look into the whole matter, he told the boy to replace the furniture. Then Bill departed for the fields.

During the day Bill was too busy to give much thought to it all, but in the evening he took Kim and Bunty up the limestone steps to the pool in the prairie beyond. He sat on a rock watching the spray from the tumbling channels of water as they hurtled down the waterfall. So it was well known that there was trouble at the Middle Bungalow. It was inconceivable that Brazier did not also know, so why had he not mentioned it? With the various rumours that he had heard before he left Dambatenne, could Brazier even be involved? A ghastly thought that he immediately dismissed. Bill also dismissed any question of the supernatural – no self-respecting ghost would deign to move furniture about and take washing off the clothes line. So it must be someone on the estate with connections with the Bungalow. There was Rasu, the *vasal kutti* (sweeper) who emptied the latrines, kept the outside clean and ran errands. He also served the Big Bungalow and the Lower Bungalow and was therefore in good position to pick up all the gossip that went on. Then there was Kuppasami, the garden coolie, with whom Dolly had already become quite friendly (by sign language!) There was no horse-keeper, and the only other coolies involved were the beef box coolie and the post coolie, but these had little real contact. Bill decided that the person interfering was either the *vasal kutti* or the garden coolie; he would start with the former.

Next morning Bill was in his office when Rasu arrived. Apart from a normal greeting Bill said nothing to him at that stage, but, as Rasu was leaving, Bill handed him four annas and a note to the Teamaker at the factory, which Bill asked him, as a favour, to deliver. Bill would like Rasu to come and confirm that it had been delivered; he would be in the plucking field above the Bungalow later in the morning.

When Rasu returned, Bill walked down to the Bungalow with Rasu following discreetly behind. Bill asked about his family and the arrangements for his caste (all *vasal kutties* were of the Pariah caste). How long had he been working at his present job? 'Only since the Sinna Dore arrived,' he said.

'Oh,' said Bill. 'Then who was working before you came?'

'My uncle, Arasan. but he could not get along with Kuppasami.'

'Why not?' asked Bill.

'Well, Kuppasami is *Peria Sathi* (high caste), and thinks a lot of himself. His mother lives in the lines on the top Division; his father was the head boy at the Big Bungalow, so a place had to be found for Kuppasami on the estate. He always thinks that he should be a house boy.'

'Thank you, Rasu,' said Bill. 'Kuppasami is a good garden coolie. I should also be sorry to lose you, so I hope that you will be able to get on with Kuppasami better than your uncle did.'

Putting his hands palmwise, and with a slight bow, Rasu departed saying, '*Amma*, Sinna Dore, *athi seria varuven*.' (That will be all right.)

Next morning, Bill called back at the Bungalow when he was least expected. Kuppasami was squatting in his hunkers outside the back verandah while Banda and Velu were unpacking the beef box that had just arrived.

When Kuppasami saw the Sinna Dore he jumped up, rather sheepishly, and made for the tool shed. Bill called him back.

'Kuppasami,' said Bill, 'I am just going to have coffee with the Doresani, but afterwards I would like to come round the garden with you. I will be out again soon.'

The garden coolie's face lit up at the Dores's interest. '*Amma, Dore, nan irrukreven.*' (Yes, master, I will be there.)

As they walked the avenue of eucalyptus, Bill said, 'You keep the garden well, Kuppasami. The Doresani is pleased with what you do. How long have you been here?'

The coolie's face brightened at the compliment. 'A long time, Dore, perhaps ten or fifteen years; I cannot remember.'

'Then you must have seen many changes, with sinna dores and their boys coming and going. How do you feel about that?'

Kuppasami did not reply at once. After a pause he said, 'My father was Head Boy at the Big Bungalow for thirty years and served only one Peria Dore. He would never allow anything to be changed. I love this garden and the Bungalow; it grieves me to see things altered by the boys, as they come and go.'

'It does not seem to me that the garden had been altered much over the years', said Bill. 'The Doresani was telling me, only yesterday, how mature everything is.'

'Yes, Dore – but the Bungalow . . .'

A sternness came into Bill's eyes. 'Kuppasami, your work is in the garden. When did you last go into the Bungalow?'

'A previous Dore used to call me in,' he replied lamely.

'What is that attached to your belt?' asked Bill.

'My keys, Dore,' replied Kuppasami.

'Show them to me.'

Reluctantly, the coolie undid the keys from his waist.

'This looks like one from the back door of the Bungalow,' said Bill. 'How did you get it?'

'Yes, Dore; one of the keys was lost by a previous Sinna Dore, and a new one was cut. I found this in the garden and kept it.'

'Then you should have returned it,' said Bill, severely, 'give it to me.'

Bill remained silent for a while. When he spoke, the tone of his voice was more in pity than in anger.

'Kuppasami, I believe that you have fantasies[2] about your work at this

---

[2] When discussing this episode with Bill some years later, I asked him what Tamil phrase he used for this expression. He did not think that there could have been an exact translation. He probably used: 'To replace in the mind a truth by an untruth.' How succinct our language can be!

place. You think that you have a right to interfere with the duties of the house boys and even to enter the Dore's Bungalow. You see some sort of duty to impose practices from the past which now no longer apply. These fantasies must stop at once. Your work here will be in the garden, and on such other work as I may give you. If you ever again exceed this you will be sent back to work in the fields. Do you understand?'

'Yes, Dore.' The expression on Kuppasami's face portrayed that Bill's words had gone home.

It was nearly lunchtime when Bill re-entered the Bungalow. Dolly was there to greet him. 'What were you talking to Kuppasami about for such a long time this morning?' she enquired.

'Oh! this and that. He wants a new hose pipe extension down to the bottom of the garden.'

'What a good idea,' said Dolly.

As 1938 progressed, the news from Home became more disquieting. Although Europe seemed a long way away, the fortunes of the British colonies were inextricably linked. Indeed, these thinly protected and widely dispersed prizes might well be an early target for aggression. Trincomalee was a naval port of vital importance as an eastern base. The port was well equipped with six-inch coast defence guns, but what of the surrounding land from which an assault could be made?

In August, before the Munich Agreement, the decision was taken to mobilise the Ceylon Planters' Rifle Corps, with the specific task to provide ground support to the naval coast defence.

The news caught everyone by surprise. Few had realised the speed with which events could happen in the event of war; there was a general feeling that everyone was going to play toy soldiers, but orders were orders. As Bill packed his equipment, he was concerned about leaving Dolly on the estate by herself. He gave Banda strict instructions that he was not to leave the estate while he was away and that he was to look after his mistress. There was little that Dolly could do – she had no car and could not drive even if she had one.

Bill had been promoted to Corporal and had his own section of young planters. He found himself on a hill overlooking the harbour: a wonderful vantage point from which to cover anyone landing at the port, but who would make such a frontal attack? he wondered. Their orders were to dig in; all plant growth was to be set aside for later covering of the spoil which was to be thrown forward of the trenches. The plant growth was then to be replaced. Bill had several questions to ask the platoon commander when the came round. Where were the other sections located, and how did their fire plan fit in with his? If their job was to provide ground support to the guns, could they not have these marked upon their maps? Where were the most likely beaches on which the enemy would land, and had they a mobile role in respect of these? The platoon commander looked at Bill with some displeasure. 'Get on with the job that you have been given, Corporal. Leave other to do theirs.'

However, that evening, all the section commanders were called together and given a better briefing.

In spite of an off-sea breeze, it was hot work digging trenches. 'What about hiring some local labour?' asked one of the section.

'For which you would, no doubt, be glad to pay,' replied Bill. 'In any case, I suppose that this is part of an exercise which may be for real later on.'

A truck came round with an urn of tea, the quality of which engendered much ribaldry. The trench was finished by nightfall. A further truck brought up two rations for those on the first guard duty and took the others back to their tents in camp. Tomorrow they would have to get a telephone rigged up, dig latrines, and get an awning over the trench. Bill was issued with two Lewis guns and four magazines of ammunition. Each man had his own .303 rifle and received forty rounds of ammo. Watch out Hitler!

Bill was right about the mobile role. Much of the next fortnight was spent surveying the possible landing beaches. If war broke out, these would have to be mined and wired, always provided that the stores were available.

The vantage point was kept permanently manned. One company had been issued with grenades and star shells. The CO made a point of visiting each section during a twenty-four-hour period and kept them up to date with the political situation. Then the news came through – Chamberlain had signed the Munich Agreement.

There was general relief all round, although many said that it was merely putting off the evil day. Soon the CPRC was packing up and returning to the estates. When they returned to their Bungalows there was a letter from the commanding officer thanking them for their service. The letter also said that the mobilisation remained in force; they were still on active service 'on indefinite leave without pay and allowances'. Later on in his service, Bill did not meet many other members of Territorial Army to whom this applied.

It did not take long to return to normal. The rugger season started in October so Bill joined the Kandy Rugger Club and was selected as the team's scrum half. The pitch was just outside Kandy with a pavilion and changing rooms. The Club's social activities took place at the Kandy Club situated on the main street in the Town, not far from the Queen's Hotel. Bill found that there was not quite the easy, pleasant atmosphere as that at the Uva Club. He put that down to the fact that he was a newcomer, from the other side of the island. It would take him time to get to know people, especially the other members of the team. The first game was against the CR & FC, the native Colombo club for whom Noel Gratien was still playing and was a tower of strength. The Kandy fly half was Jim Cotton, a good player with safe hands, but he had not the flair of Sandy Richardson. In the end Kandy lost 9-15, due largely to superior place kicking by the Colombo side. Bill was reasonably satisfied with his game, and stayed on at the Kandy club for a while to be 'matey'. It was an easy drive back to Pooprassie.

The second game, a fortnight later, was against Dikoya. Some of Bill's old pals were in that side – an additional incentive to play for a win! The evening before the match, while Bill was getting his gear ready, Dolly said that she would like to come – she also wanted to do some shopping in Kandy – so they set off early on Saturday morning. They had a snack lunch at the Kandy Club and were joined by Jim Cotton, to whom Bill introduced Dolly. They chatted about tactics for the match, before going on to the pavilion. Bill was looking forward to the game and was now more at ease with his new team.

It was a bright sunny afternoon with a welcome breeze as Kandy kicked off to gain an early advantage. The teams were well matched; it was nearly half time, with no score, when one of the Kandy three-quarters dropped a pass from Cotton. One of the Dikoya forwards pounced upon it, and, backed up by his other members of the scrum, started a furious dribble towards the Kandy line. Finding himself opposing a wall of feet, Bill fell on the ball to stop the dribble. He never knew what happened next. He must have been surrounded by the two scrums, and perhaps he was a little slow in releasing the ball. A mighty boot crashed into the back of his head.

Bill staggered to his feet, but had lost all sense of balance and direction. He fell again. The referee stopped the game and Bill was carried off.

It was some thirty hours before Bill regained consciousness. He awoke to find an attractive Sinhalese nurse peering over him.

'Where am I?' said Bill.

'In hospital,' said the girl.

'Why?'

'Because you have had a bump on the head. Go to sleep again and you will be better in the morning.' Even so, she kept a critical eye upon him for the rest of the night.

By the early morning he was feeling much better, but when he tried to sit up he had a headache and felt rather dizzy. 'Steady,' said the nurse, 'you will have to take things slowly.' She brought him some orange juice with glucose. While he was drinking it he asked what day it was. ' Monday.' she said.

'Then what happened on Saturday and Sunday? Who looked after me?'

'I did,' she said, meekly.

He looked at her in horror. 'Oh Lord!' She turned away with a smile.

Doctor Spittal was pleased with Bill's progress. The headaches would last for some time, perhaps two or three months. He should not take violent exercise during that period, but should be able to carry on normally otherwise. He would see him again in two days' time, if he still progressed, he could leave after that.

As the days went by, Bill began to realise how lucky he had been to escape greater injury from his accident. Any undue exertion gave him a sharp pain in the back of his head; it was obvious that he was going to be careful for the next month or so. Dolly noticed it too, and insisted that he take a rest in the afternoon – something that Bill found distasteful but necessary.

At the beginning of November, there was a letter from Liptons in Colombo. 'Now that you have completed five years' service with the Company, we are arranging for you to have six months' leave in England. You will remain on full salary during the whole of the period. A passage is being booked for you on P & O SS *Corfu* leaving Colombo on 4 January 1939. A further booking will be made to enable you to return to Colombo by the end of June. Perhaps you will contact our London offices with regard to that date.

'Perhaps you will confirm these arrangements. We wish you a good holiday.'

Bill had been expecting his Home leave. Even so, it was thrilling when the letter came. There was so much to do and only about six weeks to do it. He decided to trade in his Austin Tourer for a new car to be delivered at his hotel in London. He took Dolly down to Colombo to see the car dealer. After some haggling, the dealer agreed to allow Bill 1,000 rupees

(£77) against the cost of a new Austin Seven Ruby saloon at 1,755 rupees (£135) to be collected in England. The dealer would bear the cost of shipping the car out to Colombo at the end of Bill's leave. On the whole, Bill was quite pleased with the deal, especially as he was to continue to have the use of the Tourer until he sailed.

Then there was the question of leave from the CPRC. They had been warned that, in view of their mobilization, it would be necessary for all ranks to obtain leave of absence when out of the island. When he wrote, Bill also said that he would like to have an attachment to the 1st Battalion of the Rifle Brigade which was stationed at Winchester. Later he received a formal consent for his absence. The attachment had been arranged for 25 March. He would be promoted to Temporary Sergeant during the period. This would be without pay and allowances.

Dolly had written to her friends the Langdales near Chichester, asking if they could go to stay until a house was found to rent. She received a glowing and welcoming reply. Lastly, there was the welfare of Kim and Bunty. Poor Kim had not been well of late – the vet diagnosed an internal growth which was causing her much pain. After much thought, Bill decided to have her put down. A sad moment; she had been his first dog and companion, following him everywhere. Bunty was to go to a friend at Haputele – in exchange for Bill looking after his dog when he next went on leave.

Also, sadly, Velu's appointment as Dolly's houseboy had to be terminated. Bill advertised in the local paper giving him a good recommendation. He had several replies which he handed over to Velu.

Banda was more of a problem. Bill was teaching him to drive his car. He had developed into a most excellent servant and Bill hoped that he would come back after he returned from leave. Bill advertised on that basis and received no replies. It seemed that no one wanted a temporary cook appu. In the end Bill found Banda a job on a full-time basis, and had to rely on Banda's loyalty to him to return.

Over Christmas Mr and Mrs Brazier gave a party which they said was a farewell to Dolly, who had decided not to come back to Ceylon. She had bought herself a new long dress for the occasion, new shoes and handbag. Her greying hair had been specially attended. Bill was relieved to find that his dinner jacket, which had been little used since he left Kadien Lena, still fitted. They drove up to the Big Bungalow to be greeted by the head houseboy in immaculate white. Bill waited for Dolly in the hall as she left her coat in the cloakroom. Together they went in to join the other guests and to be received by the Braziers. As she entered, the others turned to look at Dolly – she gave a majestic impression of the grace and dignity of a mature woman at peace with the world and her surroundings. This was a picture of his mother that Bill would always hold and treasure. He was very proud of her.

The Braziers really became quite human – there was a lot of ribaldry about what Bill would do during his leave, together with much advice – some good, some not so good and some quite unmentionable! He learnt during the evening that his replacement on the Middle Division was to be Moore from Haputele. Bill decided to ring him with a few tips!

The day of his departure arrived; Bill's cutlery, china, and other belongings which he did not want on leave had been packed into his tin trunk and sent down to the factory. He had two suitcases and Dolly one. Dolly was quite tearful in saying goodbye to Velu; Banda was coming down to Colombo with them as his next appointment was down low country. Bill let him drive part of the way.

Commander and Mrs Neish had invited them for the night, before boarding the *Corfu* next morning, so after depositing their luggage at the Harbourmaster's house they took the car to the dealers as part of the trade-in agreement. As Bill said goodbye to Banda ('Mind you come back to me Banda – I am grateful for what you have done for me. I do not want to lose you.' 'Yes, Master. Thank you. I will come back when you return.') Bill pressed a month's wages into his hand. 'Goodbye, Banda,' he said.

Next morning Bill and Dolly were seen off by the Neishes and Evelyn Carter from Haputele, who happened to be in Colombo. They boarded in style from the Port Commission launch, flying the Harbourmaster's flag, and attended by liveried boatmen. It seemed a long time since he arrived at the jetty to be met by Leicester. Now he was a young man of twenty-four, with six months on full pay before him to enjoy himself. Little wonder that Bill was excited.

3

ENGLAND, OH MY ENGLAND!

First impressions on boarding an ocean-going liner are always confusing – SS *Corfu* was no exception. The first thing was to find Dolly's cabin and then his own. The trouble was that everyone else was trying to do the same. Then there were the dining saloon, the lounge, the sports deck and finally, and most importantly, the bar. This achieved, Bill and Dolly settled down to a large lime and soda. There was a long blast on the ship's siren – they were off. First stop Bombay.

The sea was as calm as a millpond as they steamed northwards. That night, after dinner, the ship's band played by the small square of a dance floor bordered by hidden lights. Later on, as the passengers got to know each other, there would be a danger of being pushed of the edge, but on the first night there was an air of awkwardness, soon to evaporate. From a Paul Jones, Bill found himself dancing with a Jane Elwes, and later a Pam Trenchard, the latter being married to an army officer who had been stationed in Colombo. Bill's table soon filled up with a Mr and Mrs Jakeman and a couple named Byrde.

While she was at Dambatenne, Dolly had become friendly with a Mrs Gibson who was the widow of a planter who had owned an estate near Haputele, where Mrs Gibson frequently attended church. She had often asked them back for lunch after the service, when she and Dolly found mutual interests. In addition to owning a large house at Haputele, Mrs 'Gibby', as she was known, also owned a property at Bombay, and it so happened that she was there when *Corfu* was in port. She was waiting to meet them when they disembarked. A taxi took them to the Taj Mahal hotel for coffee. Learning that it was their first visit to India, she proudly showed them round the hotel with its magnificent corridors and graceful rooms. After coffee and the tour of the hotel, Mrs Gibby left them with an after-dinner invitation to a concert at the Yacht Club that evening.

After dinner on board, Bill and Dolly caught a taxi to the Club where they were escorted across the lawn to their seats on a dais. They were treated to a fine display by the military band of the Lancashire Regiment resplendent in their white tropical uniforms. As they marched and counter-marched to the music of the band, interspersed with the thunder of the bugles, Bill felt a strange tingling down his spine. So this was at the heart of the British Empire with all its tradition and pomp. He was proud to be a part of it. When the performance was over they walked slowly back over the lawn; the sweet smell of frangipani invaded the air and the 'blue grass', imported from Australia, trod softly under their feet. Then they made their way back to the boat. What a perfect evening!

Next morning Mrs Gibby called for them before taking them back to her house. Later she took them to the officers' bathing beach where Bill revelled in a long cool dip.

There were several new faces when they got back on board, including a number of army officers. Bill attended a sports meeting and found himself elected Secretary with Carson Hyde as Treasurer and Miss Joan Kerr as Ladies' Member. There was to be some kind of event every evening until they reached Port Said. Everyone seemed happy to be going Home to England, and there was a relaxed atmosphere.

A party went ashore at Aden and there was a fancy-dress dance while passing through the Red Sea. Bill won a prize dressed as a Filipino girl with a couple of balloons under his vest and a straw skirt made from the covers of beer bottles. He and Dolly both made some purchases at Simon Artz at Port Said, and there was a light-hearted session at the Eastern Exchange in the evening. Calls at Malta and Tangier followed.

As they entered the Bay of Biscay one evening there was a call over the tannoy from the Captain. Everyone dashed up onto the boat deck to see that *Corfu* was passing the British Fleet off Portugal. They counted the lights of thirty ships, making an impressive sight.

Corfu arrived at Tilbury on 27 January. Everyone gathered in the foyer. People were tripping over each other's luggage to say goodbye. Joan Kerr came up to Bill and he gave her a kiss. They exchanged addresses and promised to write, but Bill doubted whether they would follow things up. She was two or three years older than him and rather urban in outlook.

Bill and Dolly caught the train to Victoria and then went by taxi to the Regent Palace Hotel, a vast place in which one feels quite lost. Bill rang the Austin agents, but his car would not be ready until Tuesday so they would have to stay in London. On Sunday they went down to Watford by train to see the Russwurms. All very jolly; Ann, whom Dolly had brought to Christ's Hospital, and who had spun cartwheels down the gym, seemed very well and liked the moonstone ring that Dolly brought her and Bill's leather handbag from Tangier. Ann had been working in pantomime: 'Babes in the Wood'. On Monday Bill went to see a specialist, Dr Risien Russell, about his rugby accident, as he was still having quite severe headaches. The doctor gave him a very through examination, particularly his reflexes. He enquired what he had been doing since the event and how often the headaches occurred. Bill was rather non-committal – he had been working on the principle that he should try and forget about it. In the end, Russell said that provided Bill did not overdo himself in the meantime, he should recover completely. A great relief.

The following day Bill went round to the Austin showrooms in Regent Street to take over his car. It was black with a red line outside and brown interior. It looked nice and had the smell of new paint and leather. The dealer drove him round to the Regent Palace where Dolly was waiting with what seemed a great deal of luggage for a small car. After Bill had settled the bills, the three of them piled in. The dealer drove them on to the Great West Road where Bill took over. Goodbye to London, thank goodness; at least for the time being.

While he was still at Pooprassie Bill had said that he would like to go with a party to Switzerland for the winter sports. He wasn't sure what he should do about it, so Dolly wrote to Aunt Jane at the Homestead. Jane replied that she herself could not recommend a party, but that her friend, a VAD from the First World War by the name of Lilian Killby, who lived at Bagshot, in Surrey, used to take parties over. She had spoken to her and she would be delighted to help. Bill had spoken to Lilian from London. She was expecting them for tea.

Although he kept the new car at 30 mph to run it in, they were in Bagshot by lunchtime and found the Cricketers Inn; a pleasant spot for a leisurely meal. They found Lilian's house not far from Bagshot church, where they were welcomed by Lilian and her mother. Lilian, a large jovial woman with an attractive smile and rather masculine laugh, was delighted to talk about her wartime experiences with Jane and later, of her visits to the Homestead, which, of course, Bill knew so well. It was sometime before Dolly could bring Lilian round to the question of winter sports.

It seemed that a party was being arranged by the Alpine Ski Club on 11 February – ten days' time. Two young girls, daughters of the local solicitor, were in the party. They would be glad to have another recruit! Betty Francis, the younger of the two girls, was coming round to tea to explain the details.

There was a knock at the door, and, without waiting for a reply, Betty walked in. She had natural fair hair swept back over her forehead above an open pleasant face with sparkling greenish-blue eyes. She was obviously at home with the Killbys. Lilian introduced firstly Dolly, and Bill noticed that, although Betty had sat down, she got up to go over and shake hands. To Bill she gave a smile and a nod in reply to his.

It seemed that the party was going to Austria – not Switzerland as originally planned, as there had been some difficulty over accommodation. A chalet had been booked at Seefeld, in the Austrian Alps between Munich and Innsbruck. The tour organiser on behalf of the Alpine Ski Club was Irene Kingsley-Lark, who would herself be a member of the party, which numbered about twelve. Betty gave Bill the address to write to. He would need to take ski clothes, but other things could be hired out there. They were all meeting at Victoria Station at 10 a.m. on 12 February – she so hoped that Bill could join them, especially as there were more girls than men! As they chatted over tea, Bill liked the quiet and efficient way in which Betty entered into the conversation, and yet there was an attractive femininity about her. It seemed that she came from quite a large family, well known in Bagshot. She was a friend of John, Lilian's nephew. Having left school, she was working in her father's solicitor's office. He put her at about eighteen or nineteen. She asked Bill about Ceylon – he found himself talking about it in a rather jovial offhand manner. Afterwards, he thought that this did not quite sound like him. He wondered why.

As she left, Betty shook hands with Dolly, turning to Bill. 'I hope that we shall meet again – at Victoria Station on the 12th!' She closed the door behind her.

'They are such a nice family,' said Lilian, before returning to the subject of Aunt Jane and the Homestead.

As they drove down to Liphook to visit the Isgars, Dolly turned to Bill, saying, 'What do you feel about going to Austria instead of Switzerland? I suppose that you will be travelling through Germany?' She did not say so, but the events of the Munich Crisis the previous September were still on her mind. Then, apart from Prime Minister Chamberlain's intervention, and his subsequent 'Peace in our Time' speech, the two countries would have been at war.

'Oh!' said Bill, 'it all sounds rather exciting and we can always nip out quickly, if there is any trouble.'

Dolly said no more.

Driving up to Compton House and the Langdales, the English countryside raised a thrill in Bill's breast. As they approached the South Downs the chalk hills, clothed in tight downland turf with sheep nibbling nonchalantly at the sward, fascinated him – even in the dark of winter. They passed the gates to Up Park where, as a child long ago, he had been to decorous parties from Compton House. As they reached the village, the house appeared in front of them flanked by high stone walls. 'Drive round to the Coach House,' said Dolly, and, sure enough Avice and Kathleen emerged from the kitchen door. It must have been seven years since he had last seen them. Tall and elegant, the two spinster ladies walked out to greet them like a page from a Victorian novel. Both wore long, ankle-length skirts with black crochet blouses. Their hair was plaited on the top of their heads and their long pendant gold earrings swayed as they walked. Kathleen, particularly, although the younger of the two, appeared to be stiffly clad in a bodice and high-boned lace collar at her neck. For all this, their welcome was none the less sincere. Dolly received embraces like the lifelong friend she was. To his embarrassment, Bill received a childhood hug.

They were led into the drawing-room with its heavy velvet curtains drawn away from the large round Regency window showing the neatly rolled lawn beyond. The room was scented with freesias and giant violets from the hothouse. There were gaily coloured azaleas growing in pots on the plant stands by the window. There was a roaring log fire in the grate surrounded by a huge brass fender. Such a scene made Bill catch his breath – the contrast with his life on Dambatenne was just too great.

Avice brought in the tea herself, saying that Alice would be back soon. She poured from a Georgian silver teapot into elegantly decorated porcelain china cups and there were crumpets and Dundee cake. Remembering the pitcher orchids that they had sent him in hospital in London, he was reassured of their welfare and he was later taken round the hothouses with their steamy dampness.

That night Bill sank gratefully into the feather bed with thick linen sheets and a swansdown eiderdown. This was the life while it lasted, he reflected, and he might as well enjoy the surroundings while he was here, but they were not part of him and he was eagerly anticipating his trip to Austria.

The week passed quickly. He contacted the Alpine Ski Club who were expecting to hear from him. He would need a visa to his passport which should be sent to the Passport Office without delay. Dolly would stay at Compton while he was away, but they would rent a house for the rest of his leave. One was advertised at Middleton on Sea in Sussex and Dolly would fix that up while he was in Austria. In the evening they played bridge or looked at old photos of Dolly's childhood. Avice proudly produced some mulberry wine made from trees in the garden.

During the day Bill would go for long walks up Telegraph Hill, from which he could see the coast near Selsey. Returning through the woods on the Compton Estate, there was so much game which he inadvertently disturbed that he feared an encounter with a game-keeper, but none occurred. On one sheltered corner he found a few early primroses and a dog violet.

Soon the time came for his departure; Avice and Kathleen had been kindness itself and Dolly was in good hands. He kissed them and they waved until the Austin Seven was out of sight.

And now for AUSTRIA he thought – a little tingling ran down his spine. He found Membury's Garage at Hammersmith where he left the car, paying in advance for a fortnight. Thence on the Underground to Green Park station, lugging his suitcase, where he got a taxi to the Overseas Club. He had arranged to meet Irene Kingsley-Lark at the Alpine Ski Club and found her most pleasant and helpful. She was joining the party and gave him details of the other members, also suggesting the clothes that he should buy. These, he bought in Regent Street before returning for tea at the Club.

Next morning, 12 February, Bill sat anxiously on the bus as it made its way slowly through heavy traffic to Victoria Station. Would he be in time? In the event, Irene and Betty Francis were the only other members there. Both greeted him with a smile and a handshake, but disappeared soon afterwards, leaving Bill to explore the station. He bought some tobacco for his pipe (he had had to be careful about smoking at Compton) and watched the trains as they came in teeming with commuters. There was the smell of smoke and steam everywhere, the drivers and firemen sweating and black with coal dust. A train was leaving with a full head of steam, its driving wheels slipping and vibrating until they held the track once more. Bill found the activity enthralling. A shoeblack offered to shine his shoes, but he waved him away with a smile.

Irene returned with Betty and her elder sister Joan, whom Bill judged to be about his own age. Joan was neatly dressed in a tweed suit with her light-coloured hair tied in a bun behind. He learned later that she was a qualified solicitor. Other members were arriving: Nesta Harris, Douglas McClean, Noel Tite and Victor Thorpe. Irene said that others would be joining on the boat at Dover.

As so often happens at the outset, the men sat together on the train. Bill took an instant liking to Douglas who was 'on loan' to the RAF by the New Zealand Government. He was using his leave to come on the expedition and obviously intended to enjoy himself. Of medium build with blue eyes that were full of laughter, his enthusiasm was infectious. He had straight hair with a quiff in front which frequently seemed to get in the way, and which he removed with a shake of the head. Bill wondered who would be the first to imitate him and what he did with his hair when he was flying! Douglas also seemed to have an acute sense of the ridiculous and had already produced his RAF pass to the ticket collector instead of his train ticket, waiting to see if the poor man would notice. Bill guessed, however, that there was a more serious side to him, which he was later to find out.

Although also a New Zealander, Noel was a very different character. At first he hardly entered into the conversation, but gradually he revealed that he was at London University reading political science. It seemed that he was going to write some sort of paper about attitudes in Austria after the Munich crisis, and to enjoy the winter sports at the same time. Tall, dark and bespectacled, it seemed that he might provide the brains of the party.

Victor Thorpe was a typical English lad in his early twenties, who was entering his father's business, the nature of which he did not declare. He had a pleasant open face without being over-burdened with character. His parents were paying for his holiday, and he did not seem to be short of money.

By the time they reached Dover, the four men either knew, or had surmised, a great deal about each other.

Irene was on the platform with their boarding passes for the boat, and had also found a trolley for their luggage. The girls also piled theirs on, so that amidst some ribaldry, Bill and Douglas found themselves responsible for the porterage. However, this served to relax the atmosphere. On board they were soon under way with the men migrating to the bar for beer and sandwiches. A girl strolled up to them. 'I'm Elizabeth Goff,' she said, 'I'm in your party.' After a slight pause caused by the surprise of this self-introduction, Bill and Douglas got off their seats and shook hands, Bill feeling a little guilty that she had felt it necessary to make the first advance. While Douglas was engaging her in conversation, Bill eyed her discreetly. She had an open rounded face with freckles and brown hair and eyes and spoke with easy self-confidence in a pleasant society voice.

Irene joined them with the suggestion that they should all have tea in the lounge where Joan and Betty Francis were already sitting with Nesta Harris,

Janet Domain, an older lady of about fifty, appeared to be chaperon to Jane Barker who would join them later.

It was dark when they reached Ostend. At the Passport Office they met their first spot of bother. It appeared that those with tickets through to Germany were allowed through, but for some reason or other Irene had only bought tickets as far as the border at Heidengrath, expecting to buy the onward tickets at that point. The officials insisted that this proved that they were not staying in Belgium. Eventually Janet came to their rescue by producing a hotel reservation in Seefeld. The official did not question reservations for the remainder.

Once through customs, they made for the carriages that had been reserved for them. Third class on Belgian trains only provided wooden slatted seats, but the compartments were clean and reasonably comfortable. Bill found himself with Joan and Betty Francis, Elizabeth Goff, Victor Thorpe and Nesta Harris. Most of the talk was about what they were going to do at Seefeld. They were to have a chalet with dormitories for the girls and boys. Rather primitive, but great fun. There were several night clubs in the town and the Alpine Ski Club had booked a special trip down to Munich for the Fasching Festival.

Irene came into the carriage with forms to fill about the amount of English money being taken into Germany. It seemed that one was not allowed to bring out more than one took in – an event that seemed highly unlikely! Then they searched for pencils and paper to play 'consequences' which did much to dispel any shyness that might still remain. One of the better efforts:
Neville Chamberlain
met
Ginger Rogers
in the
Flea Market.
He said to her
'My piece be with you.'
She said to him
'Never on Sunday.'
The consequence was that they both joined the war effort
and the world said
'Better to live and fight another day.'

As the rounds became more questionable there was so much laughter that someone came in from the next carriage to find out what was going on!

It was dinner time, so they elbowed their way along to the dining car. Bill sat at a table for two with Joan Francis and found her most charming and talkative. They had an excellent meal over a bottle of Graves. Joan did not give the slightest impression that she was a solicitor, an omission for which Bill was truly grateful.

Suddenly there were shouts in the corridor; they had arrived at the Belgian/German border where the dining car was to be taken off the train. They had to return to the carriages. Bill hurriedly paid the waiter.

'Thank you, sir,' he said in English. 'Now you go on to Hitler, but we stay here!'

4

YOU GO ON TO HITLER —
BUT WE STAY HERE!

As they peered out of the carriage window at Heidengrath they saw several military-looking figures strutting about the platform. A short journey to the border town of Aachen, where there was a long wait while customs officials in high peaked hats examined their luggage, which was opened and somewhat roughly handled. They seemed to be looking for cigarettes and chocolate, some of which was 'confiscated'. They emptied, or 'upset', Victor's tobacco on the floor which annoyed him. Another man entered their carriage to examine their magazines and papers . . .

The waiter's remark had not gone entirely unnoticed in Bill's mind; suddenly he remembered the consequences. He had gathered them up after they had finished playing and stuffed them in his pocket! There were some pretty rude remarks about Hitler in them; one about Hitler and Goebbels being found in the 'all in all' in the swimming pool at Bechesgarten being particularly potent. There was a gentleman in uniform on the platform staring at him through the window, which didn't exactly help.

The official came to him, picked up his copy of *Punch* and thumbed through the pages, then to *Lilliput* which had a number of cartoons depicting both Hitler and Mussolini. He pointed to an advertisement for Fort Dunlop tyres on the front cover, asked in broken English what it meant, and passed on . . . Bill heaved a sigh of relief!

Conversation was a little dispirited after the customs episode; it was not long before the train was due to leave, that they suddenly realised that they were supposed to have their money examined and checked on the platform outside. Result: a general stampede. They had to empty out their purses and show their passports, just managing to get back in time. A rather tiring three hours followed, on the way to Cologne, where they were to change. Most spent the time dozing or reading until they arrived at midnight.

In Cologne they had an hour to wait. There was a *weinstube* with flashing green and yellow lights which they explored, ordering beer for the men and wine for the girls. The *stube* was packed with people and very hot and smoky. After some time, Bill suggested to Betty, who was sitting next to him, that they should walk up and down the platform for a while. It was the first time that he had had an opportunity to talk to her on her own. They were both rather tired, but the air was fresher than inside. Bill hoped that he had been good company. It was 1 a.m. by the time that their train arrived. The party had to divide as Bill, Noel and Douglas had booked sleepers while the rest had to spend the night on the wooden slatted seats. Bill and Noel found themselves in a three-berth compartment with a German boy from Berlin. The sleeper was terribly small; only one person could undress at once, but as soon as they had sorted themselves out they were asleep.

It was all change at Munich, where they went on a quick tour of the city. Bill was enthralled by the magnificent old fifteenth- and sixteenth-century buildings which towered above the wide streets. There was a sombre splendour which portrayed the beauty of earlier civilisations. Now there were SS officers in black uniforms with swastikas on their arms; black swastikas on white and red backgrounds hanging from nearly all the buildings. Bill surmised that it was unpatriotic not to do so. There were huge posters: '*Juden sind verboten*' (Jews are forbidden) and others, '*Deutschland über Alles*'. For all that there was an air of gaiety about the place, and banners and decorations were already being erected for Fasching.

After the tour, they raced back to the Town Hall to hear the hour strike and see the figures dance. Firstly the bells rang out with a wonderful depth of sound and tone, the Great Bell reverberating and echoing down the main street. Then some dancers appeared, dressed in medieval attire, to move gracefully around the clock face. With the bells still playing, two horsemen then appeared to pass each other once before turning for the assault. With their lances raised, one was struck on the breastplate and knocked back on his horse. A fine display.

Returning to Munich Station, Bill again sat next to Betty during lunch, and apologised for his somewhat sombre company at Cologne station. 'Yes,' she replied quietly, 'We were both rather exhausted, but I was grateful for your suggestion that we should get some fresh air.' She continued with a laugh, 'What made me absolutely livid was that you should have had a sleeper, while we mere girls should have had to sleep on the hard seats! Then, to cap it all, we had to turn out at Ulm and move to *your* part of the train!' Bill looked at her and admired this humorous and provocative defence of female prerogatives. Her manner was slightly teasing.

'What a shame, Betty, but you didn't miss much. There was hardly room to move in those cabins,' was his rather lame reply.

Finally they climbed on a packed train to Seefeld, where they arrived at 10 p.m. A guide took them to their chalet where the *Hausfrau* was waiting for them. They were all asleep in no time.

Next morning they awoke to find a blanket of snow covering the ground. There was brilliant sunshine which was reflected from the powdery snow like diamonds, dancing and glistening to their tread. After a quick breakfast of rolls, butter and jam with a large cup of milky coffee, they all departed up to the nursery slopes in their ski clothes. There was an attendant at the entrance where one hired skis and sticks. This was an entirely new experience for Bill, who felt rather a rabbit. However, only Elizabeth, Douglas and Victor had previous experience. They did little more than get the feel of the skis and learn to keep their balance on the simple slopes. Luckily the slope levelled out at the bottom so that they did not have to worry about stopping! All the instructor's orders were in German, so they had to learn them quickly. There was a deal of ribaldry about what had been said by the instructor – it was left to Elizabeth both to translate and to demonstrate; soon they were taking more notice of her than of the official.

At five hundred metres altitude, the air at Seefeld was crisp and clear. The sunshine continued unabated so Bill, Nesta and Betty Vaughan made for the Wetterstein skating rink. It was six years since Bill last took to the ice, at Warnham Lake at Horsham. Nesta and Betty V were both accomplished, but, after a few rounds by himself, Bill found his feet once more. By the end of the of the afternoon the three were holding hands and gliding round the rink in time to the music. Betty was a tall, elegant girl, whom Bill guessed was rather conscious of her height. Nesta, on the other hand, was petite and graceful, but gave a rather 'brittle' impression. How different they all were! Later, Bill found that there was a certain 'wiriness' about Nesta that belied her looks.

The afternoon was a great success. Bill loved the feeling of freedom and co-ordination that the ice gave him and there was a sense of satisfaction that his old prowess had returned. He looked forward to many more visits to the Wetterstein.

In the evening, after supper, they all arranged to visit the Hohemunde Café. The men wore lounge suits and the girls evening dress. Their party of fourteen monopolised a corner to themselves, to be entertained by the band playing soft music. The men took turns to order jugs of Rheinwein and Apfelsaft amidst much spontaneity. Then the band struck up the first dance and they looked around for partners. Bill went round the table to ask Betty to dance 'to make amends for having had a sleeper'. It was a particularly German type of two-step to which Betty responded; they both were out of breath and flushed of cheek by the end. Douglas was dancing with Elizabeth and Victor with Irene, but Noel was 'sitting on his hands'. It was a pity that there were not more men.

Bill was really enjoying himself, and during the evening danced with each of the girls, except Betty Vaughan and Janet Baron, who were too tall! The Rheinwein and Apfelsaft flowed; by the end, if they were not dancing, they were singing and stamping in time to the music, but no one seemed to mind. They stood the band a drink each, and left late, to find the snow-clad path to the chalet.

The next day was rough. The skies were leaden and full of snow; the wind blew. They were advised not to go out too far, and spent most of the time writing letters, playing cards or reading. No one was really sorry; they had had a hectic time since leaving England. The *Hausfrau* brought them coffee and Bill found himself next to Noel, with whom, up till then, he had had little contact after the train journey to Dover.

'I hope that you are enjoying yourself,' said Bill. 'Last night was rather rowdy!'

'Oh! yes,' was the reply. 'That was what I expected.'

Bill was about to say, 'Then why didn't you join in?' but bit his tongue. After a pause, Noel continued:

'It is all a little unreal really; only four months ago, Austria was on the point of war with Czechoslovakia, now, outside this little environment of ours, people are holding their breath.'

'Was it really as close as that?' said Bill. 'Tell me more about Munich.'

'Well,' said Noel, 'as you know, Hitler was on the point of marching on the Czechs last summer, possibly as a prelude to invading Poland. But Britain had a treaty with the Czechs; Hitler's intervention could, therefore, have brought Britain into war with Germany. Chamberlain came to Munich last September to warn Hitler that Britain would do just that – the question was whether Chamberlain was bluffing. That we shall never know. In fact they reached a compromise – the Czechs would surrender the Sudetenland, which is an area not far north of here, to Germany, in return for a diplomatic agreement. The Sudeten extends into the belly of Germany and I suppose that there was some vague territorial claim.'

'And is that what has happened?' said Bill.

'That is what interests me,' said Noel. 'I hear that not much has happened; it almost looks as though Hitler is waiting for something else. I am planning to go on there later.'

'Are you, by jove?' said Bill. 'But I thought that you didn't speak German.'

'Not in public; one learns more by listening than by speaking. I have relatives in the Sudetenland – my mother was Czech.'

Bill looked at him in admiration. 'Then if Hitler decides to ignore the Agreement and goes into Czechoslovakia, we may all find ourselves in a jug,' he said with a smile.

'I doubt whether things would happen as quickly as that,' said Noel. 'Most organisations with people abroad are in touch with the Foreign Office. I know that Irene has been in touch with London.'

Bill was silent for a while. He had given very little thought to the effect of the Munich crisis upon their holiday, and he doubted whether the others had either. He decided not the raise the subject with them. Then he turned again to Noel.

'What can you tell me about the anti-Jewish posters that we saw in Munich, and again up here?'

'Yes, they are not very nice,' said Noel. 'I suppose that the anti-Jewish feeling is part of the build-up of German nationalism that has been fanned by Hitler and Goebbels. Last November there were huge demonstrations against the Jews, including some in Munich, which were called *Kristallnacht*, 'the night of broken glass'. Since then, thousands of Jews have been arrested for so-called anti-national activities.'

'What will happen to them?' asked Bill.

'Goodness knows. Israel is a small country with only limited opportunity. Many Jews have sought employment abroad and have been the victims of their undoubted entrepreneurial skills. At the expense of the countries' own nationals! On the other hand, if there is a war, Germany will need cheap labour that can be directed to unskilled or dangerous jobs. I suspect that Hitler will keep them locked up for the time being, to see what happens.'

'And to some extent you are also here to see what is happening,' said Bill. 'I wish you the best of luck!'

At that moment there were shouts from down below, coming from the path leading up to their chalet. Looking over the balcony, Bill could see a snowball fight in progress between Douglas and Victor, on the one hand, and Elizabeth and Nesta on the other. 'Excuse me,' said Bill, 'I must go and see fair play.'

That evening there was a Ski Ball at the Hohemunde which they attended in *Ski Kleider*. This time, the girls ordered the Graves and the Apfelsaft and they occupied their usual corner. During a Paul Jones (which was actually introduced as such!) Bill found himself with a rather large and copious Austrian woman who insisted on talking to him very rapidly in German. However, he managed to say '*Ja*' and '*Nein*' more or less in the right place and even sometimes '*Jawohl, das ist rechtig,*' which seemed to please her. Then there was a sort of musical hats. A ski cap was passed round until the music stopped, the wearer and his partner then being 'out'. Bill was dancing with Nesta – they were in the last three couples. Lastly, two men held a ski stick horizontally above the floor. One had to dance underneath without touching the stick. It was lowered every round, the winners having to crawl under on their hands and tummies. Bill was dancing with Irene, who, poor girl, with her girth, knocked the stick quite early.

Next morning was fine and clear. They were all on the ski slopes early, the beginners becoming more confident. Douglas and Elizabeth disappeared to challenge the training slope, returning an hour later, Douglas carrying the stub of a broken stick.

In the afternoon Bill went with Joan and Betty Francis to the Wetterstein. They were both learning to skate and Bill found himself helping them. In the end, all three, with Bill in the middle, were striding out quite well to the music, which gave them a wonderful sense of time and rhythm. Then they sat on a seat by the rink, chatting away merrily with anecdotes about the remainder of the party, and how things were at home in England. Before they left Bill asked Betty to come with him to the Sports Café that evening. It was to be a Ski Ball in *Abend Kleider*, so that they would have to 'dress up'. After a slight hesitation, Betty accepted and arranged to be ready at about 9 p.m. This was something of a departure from events so far, as the party had tended to stick together. Bill was not quite sure how their absence would be received, as there were so few men. However, another unattached male – Ivan Churcher – had just joined them from England, and Bill thought that he had 'done his duty' so far. Betty joined him with a black satin shawl, and looking very nice indeed. They both felt a little self-conscious in front of the others who were still in their *Ski Kleider*.

The Sports Café was an altogether more 'select' affair. They were ushered to their table (which Bill had surreptitiously slipped away to book beforehand) and ordered Heinkel Trocken. It suddenly struck Bill that this was really the first time that he had ever taken a girl out to dinner *à deux*. He rather guessed that the same might apply to Betty, only, of course, in reverse! Bill did not remember what was on the menu. The meal was a leisurely affair interrupted by dancing and the replenishment of wine. At first the band was muted, playing Strauss waltzes and slow two-steps. Master at the use of an Austrian fiddle, the leader, in a Tyrolean hat complete with a cock's feather, was ambivalent as a master of ceremonies. He was there to make his guests enjoy themselves. Once the dishes were cleared away the atmosphere livened. They danced and whirled to Viennese 'old-fashioned' waltzes which Bill had always loved, particularly with a good partner.

Then, a girl dressed in Tyrolean garb appeared to muted half-lighting to sing a lilting song:

> *Küss mich, bitte, bitte küss mich eh die nächste Bahn kommt,*
> *Küss mich ohne Pause.*
> *Küss mich, bitte, bitte küss mich wenn die Bahn dann ankommt*
> *Muss ich ja nach Hause,*
> *Sie macht mit geläut, sie hält nicht oh Schreck.*
> *Ich wart auf die nächste, dann muss ich weg.*
> *Küss mich, bitte, bitte küss mich eh die nächste Bahn kommt*
> *Dann muss ich nach Hause.*

Bill wasn't sure how much of the song Betty understood, but the mime was expressive. He touched her hand and noted a slight expression of embarrassment.

A photographer came to their table wishing to take their picture for a mark, and Bill asked for two copies. Then there was a gallop, the couples prancing around the floor and then, somewhat hazardously, across the middle. It was all great fun.

It was two-thirty in the morning by the time the evening ended. They walked back along the snow-covered path holding hands as they went. At the door of the chalet, Betty turned to Bill:

'Thank you Bill, I shan't forget this evening.'

To the ski slopes again next day. Elizabeth and Douglas disappeared on their own leaving the rabbits on the nursery slope once more. However, Bill found that, in common with the others, the sense of balance was coming.

They had learned to lean forward to counter the movement of the skis and to bend their knees to even out the hummocks. Leaning sideways on corners was still a problem as was the crossing of their skis. As for stopping . . .

At breakfast, Bill had asked Elizabeth to come riding with him in the afternoon. He had arranged to hire two ponies after lunch; a groom was waiting for them.

'*Hier is Fritz, und das ist Lilley. Bitte wenn werden Sie zurück kommen?*'

'*Am vier oder fünf Uhr, glaub' ich,*' said Bill.

'*Schön, gute Reise,*' said the man as he left them.

After casting a professional eye over the filly, Elizabeth chose Lilley. She spurned Bill's offer of a 'leg up' and they were soon on their way to Mosern, a village up in the mountains above Seefeld. Both ponies had spikes attached to their shoes to grip the snow. Bill found that Fritz had a good mouth and they were soon trotting along the mountain path. There was a cloudless sky; the reflections of the sun's rays on the snow was, at times, dazzling. They came upon a wide valley between the hills where the snow lay in virgin whiteness before them, the beauty of which was breathtaking. The way was now lined with towering conifers, their branches drooping and bent with the weight of snow upon them. Closely planted spruce, their straight trunks reaching upwards towards the light, swayed gently to the breeze. Everywhere was the clean, slightly resinous smell of pine. Bill warmed to the willing effort of the horse beneath him; to these wonderful surroundings, and to the company of this girl whom he had known but a week.

They came upon a little Catholic shrine built into the rock by the side of the path. A figure of the Virgin Mary gazed up at them. Bill said a silent prayer of thankfulness before urging Fritz on the way.

They reached Mosern village. The hotel stood at the head of a steep road with a magnificent view over the peaks of the mountains. They sat on the balcony in the brilliant sun drinking Apfelsaft and chatting. Elizabeth asked him about Ceylon and the horses he had had there. He told her about Mary and his experience when she was killed. She had a distant relative out there, whose name she couldn't remember, but seemed genuinely interested. 'Sometime, if I ever do a world tour, I must come there,' she said. 'Then I will look you up.' Just them, Ceylon seemed to be a very, very long way away.

On the way back, the sun was beginning to melt the snow on the branches of the trees. Splashes fell around them as if coming from nowhere, causing a 'plopping' sound. Bill had to tighten his hand on Fritz's rein to reassure him. They had a good canter on the level as they re-entered Seefeld.

Next morning, Victor did not appear for breakfast. Bill knocked at the door of his room twice before entering receiving a muted reply. Entering, he found Victor holding his stomach and groaning. 'Leave me alone,' he said.

It seemed obvious that he had picked up some sort of bug. Nesta wanted Bill to go along to the doctor without telling anyone, but Bill thought Irene ought to know, so they both went along to the local *Arzt*, where Bill had to describe Victor's symptoms. Somewhat surprisingly, the doctor came with them and Bill stayed while he examined Victor. In the end there was nothing that a day in bed and warm drinks would not cure – much to Irene's relief.

By the time that this had been dealt with, it did not seem worth going up to the ski slopes, so Bill joined Betty, Joan and Winifred Turner at the Wetterstein. By now they were all really quite competent. Bill remembered particularly gliding around with Betty to the tune of the 'Skaters Waltz'. Their movements in unison thrilled him – he wished that they could go on forever. As the music died away, they collapsed onto the seat exhausted, Betty's hair windblown and her face flushed. Bill departed in the direction of the kiosk for some hot coffee.

For the next day, Saturday, Irene had arranged the memorable trip to München for the Fasching ceremony.

Most of the population around this part of southern Germany is Catholic. In the run-up to Lent the people seem to work themselves into a passion, culminating in the magnificent eve-of-Lent Festival of Fasching. The Ski Club party left Seefeld by train in the early morning and were glad to get a carriage to themselves. Inevitably, such a gathering of young people on holiday was a noisy affair, singing songs and telling stories. The guard who came to check their tickets gave up halfway through – these English! Douglas produced a can of Bier which was passed round tenderly like a loving-cup. The three-hour journey passed quickly.

Crowds were massing when they arrived in München. They were met by Irene's German friend, Garbi, who bid them follow her, threading their way out of the station. She led them to a café (which Bill thought to be more like a *Weinstube*) where she had booked lunch. They were glad to be out of the crowds for a time. They had barely finished lunch when the procession started. Some rather 'tight' and not very pleasant young Germans tried to ouzel their way into the party, but they managed to shake them off, before departing to see what was going on outside. The procession consisted of tableaux mounted upon carts and lorries depicting political and topical characters and events both of national and local relevance. They had been led to expect these to be more extreme and military in nature than they were – could it be that southern Germany did not share the apparent ambitions of the Nazis? As the procession progressed the effigies became more and more grotesque as they swayed and gyrated in time to the music. All the local bands seemed to be there, many in picturesque Tyrolean costumes with girls dancing before them. There was even a tableau of Snow White and the Seven Dwarfs which was cheered loudly.

After the procession had passed, the group decided to split up for the time being, Elizabeth, Betty, Joan, Freda, Ivan and Bill deciding to go off on their own after having arranged to meet the others at the Regina Hotel later on. First of all was a visit to the Haus Der Deutscher Kunst which was a permanent exhibit of the national achievements carried out during the Third Reich. The building itself with large porticos and polished marble floors was impressive, but much was to come. There were miniatures of most of the great establishments constructed by the Nazis. There were replicas of Hitler Youth hostels, hospitals, railways and even Hitler's palace! The Germans seemed to be working on early motorways as there were examples of overhead road junctions, apparently invented by a young engineer of only eighteen. There were also interiors of some of the great houses and ships being built. These were lit by innumerable little electric lights in proportion to the miniatures.

There were some very powerful and explicit male statues looking down upon them from the walls of the halls, symbolising German youth and strength – something to aim at and achieve. Bill saw Elizabeth looking at them and caught her eye. 'Don't worry,' she said, 'I have two brothers!'

'What do you think of the exhibition? It all seems to be aimed at self-congratulation.'

'Yes,' she replied. 'It's all propaganda, of course, but I don't really see why not. It's all aimed at the advancement of youth and national pride. They have kept the problems like the Jewish question and the future of the Czechs in the background.' Bill found the halls rather aggressively German, but he did not advance that view.

There was then a visit to a museum of rural life which was a little less controversial, before returning to the Regina Hotel (the hotel where Chamberlain, the British prime minister, had stayed) to meet the rest of the party who were already installed at a large table waiting for them. Dinner was a splendid affair and, as usual, they danced between courses. Bill remembered dancing with each girl in the party and with Betty several times – that is when she wasn't being chased by Ivan, which, thought Bill, appeared to be unnecessarily often!

They were turned out of the Regina at about 2 a.m., so went off to find somewhere else to continue the night before their train left Munich at 5 a.m. As they walked along the main street in Munich, Bill and Victor started singing 'It's a long way to Tipperaray'. Soon the others joined in, linking arms and marching down the middle of the street. Then there was 'John Brown's Body' followed by 'O God Our Help in Ages Past', all sung with much verve and not much reverence. Such was the noise it was surprising that they did not land up in a concentration camp!

Then they found a *Weinstube*; a weird place, in which a few couples were dancing on a small central floor, with brightly coloured lights shining up through a plate-glass centrepiece, casting exotic shadows by those passing over. A man was playing a drowsy tango on a concertina in an atmosphere thick with tobacco smoke. They found a corner and ordered wine and beer. Douglas and Elizabeth started to dance.

Bill was not sure how long they had been in the *Stube*. The thickness of the air, the desultory music and the lateness of the hour had dampened their spirits. Bill hardly noticed that some Germans had joined the party, one of whom had ordered a *Stiefel* (a glass replica of a top-boot) filled with *Heles Bier*. This he toasted to everyone before taking a sip himself and passing it round urging each to do the same. Suddenly the party sprung to life again.

Bill became aware that a young German girl had come to sit next to him. She had straight hair, pulled to a bun behind her head. Her dark eyes were looking at him. He guessed that she was about twenty-one or twenty-two – a student perhaps?

'*Heil!*' she said. Then, after a pause, '*Wie findest du unser Fasching?*'

'*Ich hab'es sehr gern, danke,*' replied Bill, '*ist es immer so gut?*'

'*Ich denke doch, aber ich habe noch niemals daran sehen,*' she said.

'*So, was tun Sie hier?*' said Bill

'*Ich bin an der Universität in München. Ich heisse Heidi – wie heisst du?*'

'*William,*' said Bill, a little self-consciously.

'*Ach, mein kleiner Wilhelm!*' laughed the girl, provocatively. Bill looked at her, but could do not other than take the jibe in good part.

The concertina was playing a waltz. Heidi got up and held out her hand. '*Tanzen?*' she said. She was slim, neatly dressed in a straight-fitting dress and a white collar. She looked attractive. He found her even more attractive on the floor, where they were swinging around to 'The Blue Danube'. It seemed that the man with the concertina was making a special effort on their behalf. Heidi responded vivaciously. As the dance came to an end, Bill twirled her around on his fingers in front of him before bowing to her curtsey.

Taking his hand, Heidi led him to an alcove away from the others. As they sat down, '*Was trinken Sie?*' he asked.

'*Wein, bitte,*' she replied. While they were waiting for the wine to arrive, she lit a cigarette, after having offered one to Bill, which he declined.

'*Du bist Englisch?*' she enquired.

'*Jawohl,*' was the reply.

'*Wohin aus England?*'

Bill thought for a moment. There seemed little point in introducing the subject of Ceylon. '*Aus London,*' he said, rather lamely. He might have done well to ponder Heidi's use of the familiar '*du*' when speaking to him, but the point went unnoticed.

'*Dann was machst du hier?*'

'*Wir sind alles auf Winter Sport aus Seefeld.*'

Bill thought that it was about time that the line of questioning was reversed. '*Und du – bist du einsam hier?*' She certainly seemed to be alone. Bill was beginning to wonder why. She waved the question aside.

They danced together again, this time a slow tango. Bill felt her body relaxing against his – he did not resist.

Back in their corner once more, Bill saw her looking past him over his shoulder. He did not see the party leaving the *Stube*.

Heidi was playing with her half-empty glass, as if to make a distraction. After another look beyond him towards the door, she put her hand in his and moved closer to him.

'*Du hattest nicht nun nach Seefeld zurück gehen. Durch machen die ganze Nacht! Kommst mit mir – ich habe Einzelzimmer. Morgan konnen wir die wurdigkeiten besichtigen . . .*' Suddenly Bill realised that he was being propositioned!

At that moment, there was a thump on his shoulder and a stern voice behind him. 'Come on, William, we are all waiting for you!'

Releasing Heidi's hand Bill jumped up, but, remembering his manners, he turned to the girl. '*Unglücklich, Heidi, nun muss ich weg. Danke zur deiner guter gesellschaft!*' He bent down to kiss her hand before turning on his heel to follow Douglas to the door.

'Was she very nice?' said Douglas with a mock obsequiousness.

'That depends,' replied Bill reservedly.

'Depends on what?' pressed Douglas.

'Upon what you mean by 'nice'. Just now, I'm not sure that your definition would agree with mine!'

'Ah well,' concluded Douglas, 'you seemed to be getting on quite well together. Incidentally, it was Betty who noticed that you had not come out with us when we left.'

They caught up with the others to walk through the by now almost empty streets on the way to the railway station. Their train was almost empty, too, so they spread themselves along the wooden seats for what comfort these would provide. On reaching the chalet at Seefeld, Bill flopped unceremoniously into his bunk.

It was lunchtime when there was a knock at the door. Betty had brought soup with some rolls and butter which she had got from the caretaker. 'How's the head?' she enquired solicitously.

'Awful.'

She smiled. 'We thought that you were going to be left behind!' she said meaningfully.

'Yes, I know,' said Bill. 'Douglas told me that you had deputed him to come and rescue me.'

'Did you need rescuing?' she said, laughing.

'Well . . . Oh, let's change the subject! What are we all doing tonight?'

'I think that everyone is going to the Hohemunde,' Betty replied.

By the evening they had recovered sufficiently to settle themselves into their usual corner at the café. During the first dance, which he had with Betty, 'Are you quite exhausted after yesterday?' Bill enquired.

She didn't reply at once, but after a pause: 'I loved the procession and the exhibitions. Also the gaiety of Munich with its beautiful buildings together with the colour and attraction of the place and its people. But, really, the day was too long. If only we could have come back to Seefeld after dinner at the Regina. That last place was ghastly!' She finished speaking with such emphasis that Bill looked at her, but did not pursue the subject.

'I have not seen Noel since we got back. Do you know where he is?' said Bill.

'He told Irene that he was going to stay in Munich for a few days. He may be joining us again later. Irene was rather surprised.' Remembering his earlier conversation with Noel, Bill was able to guess what he was about.

They resumed their seats after Bill had ordered Apfelsaft for them both.

The evening continued with its usual exuberance. By now, they all felt thoroughly at home at the Hohemunde. They knew each member of the band by his first name; the leader being Kurt, who was repeatedly called upon for 'request' numbers.

While Bill was dancing with Elizabeth, the lights were dimmed as a scantily dressed girl appeared upon the platform carrying a small hand mirror and comb. After a few bars of music, she began to sing the haunting and lilting song of the Loreley in which a beautiful maiden, who sits upon a rock overlooking the Rhine, sings as she combs her golden hair. A boatman, in a small boat, becomes so enchanted that he is wrecked in the swirling waters of the gorge.

Both Bill and Elizabeth knew Heine's famous song by heart and began to join in the song. When the girl had finished, Kurt called them over and suggested that, in the encore, they should both join the platform. But Bill had a better idea – Elizabeth should be the *Jungfrau* and he would be '*Den Schiffe in kleinen Shiffe*'. As Elizabeth took up her position sitting upon a table on the stage, Bill grabbed another, overturning it for use as a boat.

Elizabeth was magnificent as she sang,

> '*Die schönste Jungfrau sitzet,*
> *Dort oben wunderbar,*
> *Ihr goldnes Geschmeide blitzet*
> *Sie kämmt ihre goldenes Haar.*'

Meanwhile Bill, with the aid of two hastily acquired dinner forks for use as paddles, became entranced and bewitched –

> '*Den Schiffe in kleinen Schiffe*
> *Ergreift es mit wilden wey,*
> *Er schaut nicht die Felsenriffe,*
> *Er schaut nur hinauf in die Hoh.*'

The climax was reached after Bill, spilling his 'ship' before the beautiful maiden, lay prostrate at her feet as accompaniment to the final lines of the song,

> '*Und dasz hat mit ihrem Singen,*
> *Die Loreley getan!*'

This piece of buffoonery was greeted with hoots and whistles. It was, perhaps just as well that the evening was well advanced!

The Ski club had ordered a round of drinks for the band on several occasions; this time Kurt saw an opportunity to return the compliment during which Bill and Elizabeth stood with him on the dais linking arms before drinking a '*Brüderschaft*'. Something that Bill was to recall, uncomfortably, in later years.

At breakfast the next morning the sun was already shining through the windows of the chalet, its rays reflecting the whiteness of the snow as dancing crystals upon the ceiling. 'A day for ski,' announced Douglas firmly. And so it was. While Elizabeth and Douglas disappeared gracefully into the distance, the others practised their stems turns and balance somewhat unrewardingly. 'Thank you,' said Betty to Bill as he pulled her out of a bank of snow on one of the corners just before Joan followed with an inelegant spreadeagle into the same obstacle. At that moment Victor arrived with a camera. 'Oh no, please!' said Betty, but he snapped them just the same avoiding Betty's snowball in the process.

The afternoon saw Joan and Betty, Nesta, Victor and Bill at the Wetterstein. After a few turns linking hands around the rink, Bill took Betty off on their own for the 'Skater's Waltz'. He felt her hands warm in his, even through their gloves. Suddenly, Bill felt an indescribable sense of exhilaration – the freshness of all around them, the mountain air, the sense of rhythm in their effortless movement, the music and the response of this girl beside him. They did not talk – it was not necessary. Anyway, he would have been afraid to impart his feelings.

It was not to last. As they left the rink Betty announced, rather hesitatingly, that Ivan had asked her out for the evening at the Sports Café.

As the others made their plans for the evening over the supper table, Bill decided that he wanted a time to himself. Everything seemed to be moving so fast with relentless pressure to plan the next enjoyment. Making his apologies, he donned his ski jacket before walking out into the crisp night air.

Uncertain at first where to go, he made his way along the snow-covered streets in Seefeld, before finding the road to Griszenbach. The moon was shining through a cloudless sky; with care, he could make his way quite easily. There was no traffic – all he had to do was to follow the wheel tracks in the snow. Two men passed him, the light from their cigarettes flickering. '*Abend*,' said one. '*Guten Abend*' replied Bill. Then he was alone.

As he walked, the incidents of the last few days came flooding through his mind. The incident with Heidi in the *Weinstube* in München still bothered him. She had seemed to be an educated, well-dressed young girl very far from his picture of a common prostitute. She was a good and practised dancer – one most unlikely to be short of her own boyfriends.

Why, then had she picked upon him? Why was she so interested in where he had come from?

Supposing he had left the party and gone with her to her apartment: what would have been the motive and was this to have been in some way sinister? He supposed that this was possible, but rather unreal. Then there was the nagging thought that she was in some sort of trouble – needing protection. No, surely not?

Then there was the possibility that he had misunderstood what she had said. No, that wouldn't do either.

He was really left with only two alternatives. Either Heidi's intent was sexual and mercenary, or she genuinely wanted to show a foreign visitor some of the beauties of the great city of Munich.

Bill had enjoyed his visit to Munich, and, indeed his meeting with Heidi had not been without its charms. However naive it might be, the latter alternative was the one by which he would choose to remember the incident!

A distant light from a car rounding a corner ahead on the mountain road brought his thoughts sharply back to the present. He stood to one side as it passed, but even so a gush of snow covered his jacket. As the vehicle rounded the next bend out of sight, a bat passed over the face of the moon, then another and another. With a smile, he remembered the bats in the cave at Bandara Eliya, when Diane Gorman had spilled the hurricane and they were surrounded by the frightened animals. Then there was the consternation upon Mrs Gorman's face when they both returned, covered from head to foot in guano!

But all that was a very long way away and he was here to enjoy himself.

After his exhilaration dancing with Betty on the Wetterstein that afternoon he had felt a vehement pang when she told him of her invitation from Ivan. He had told himself that Ivan had every right to take her out if he wanted to – and yet . . . He remembered a boy from Yorkshire in his house at Christ's Hospital, who when under pressure used to say, 'Eh laad, watch it!'

A cloud passed over the moon and it was getting colder. Alone with his thoughts, he had walked a long way – miles perhaps. As it grew darker his mind turned to that tragic German poem of the father riding through the forest with his dying son in his arms:

> 'Wer reitet so spät durch Nacht und Wind:
> Es ist den Vater mit seinen Kind'

As the Devil teases the father before taking the boy, the father's torments grow ever greater until finally,

'In seinen armen dasz Kind war tot.'

How typically German, thought Bill as a slight shudder went down his back.

He was walking faster now, his footsteps sounding crisper as the frost hardened the packed snow. He sang *'Küss Mich, bitte, bitte, bitte Küss Mich'* and incongruently, *'There will always be an England'* to himself as he walked.

On the way back to the chalet, his route took him past the Sports Café. He paused for a moment to listen to the music; the orchestra was playing the Strauss waltz 'Tales from the Vienna Woods'. As he turned away, 'Damn Ivan Churcher!' he thought.

He was glad to scramble into his bunk; there would be another day tomorrow. He was asleep before the others returned.

As the days passed, the initial excitement of the strange surroundings was replaced by their growing friendships and sense of prowess on the ski slopes and on the ice rinks. Betty Vaughan and Elizabeth Baron had become 'attached' to two young Austrians, Hans and Geppi, though, as neither of the girls could converse in German, or they in English, there was general speculation about how the girls managed. Members of the Hohemunde band had also attached themselves; Kurt was always good fun and Leo, the pianist, could speak quite good English.

One evening, with a mumbled remark 'With all these Germans about things seem to be rather crowded,' Douglas suggested to Bill and Ivan (now forgiven by Bill) that they should go on a 'pub crawl'. Starting at the piste, they dived into a bar and ordered bier.

Douglas was in great form: always full of fun, tonight he turned his attention to 'taking off' other members of the party. He revelled in his description of Geppi, who was short and fat with curly hairs bulging over the top of his vest – a sure sign of over-sexiness. Even Elizabeth did not escape his mirth: he suggested she should take a slimming course! As for Nesta . . .

Later some rather loud-mouthed Germans came. Ivan suggested that they joined the others at the Hohemunde. When they left, Bill took Betty's arm on the way back to the chalet. 'I'm not sure that I approve of these male stag nights,' she said. 'I had to dance with the 'Woolly Bear' which wasn't very pleasant! If you men must go off on your own, can you find better substitutes?'

'Point taken,' replied Bill. 'We will try and do better in future.'

Next morning Betty came into breakfast limping from a sprained ankle. Bill offered to accompany her to the local *Artz* to have it attended. While they were waiting in the surgery, they chatted happily.

'We haven't much longer here,' said Betty. 'Are you glad that you came?'

'I have enjoyed every minute,' replied Bill, 'I wish that we could all stay on — it seems such a pity to leave this lovely place.'

'What will you do with the rest of your holiday — you've another three or four months haven't you?' said Betty.

Bill was a little flattered by her interest in his future. 'Well, I have to fetch my mother from Compton, where she is staying with friends. She is arranging to rent a house for my leave. Then the Ceylon Planters' Rifle Corps has fixed up an attachment for a week in May with the 1st Battalion of the Rifle Brigade at Winchester. Then there are relations to see . . . but that won't take long. Could we meet when we get back to England?'

'Well, that's why I asked, really,' she said. 'We have all been taking photos and buying cards. Irene thought that it might be fun for us all to meet at the Alpine Ski Club after we get back to compare notes.'

'Count me in on that,' said Bill. 'It's a fine idea.'

The doctor beckoned them. His English was better than Bill's German! Luckily there was not much wrong with Betty's ankle.

While they were in the surgery together, Bill had asked Betty to come riding with him, and had later managed to book Fritz and Lilley for later in the week. Bill found himself looking forward to this; Betty had her own pet horse in England which she often hired in the holidays. Off along the Griesenbach road, both horses were eager for their heads; in fact Bill who was on Fritz, has to restrain him quite forcefully. 'They must have had a good bran mash last night!' said Betty, laughing. Near Seefeld there was some traffic which had caused the snow to melt. Bill loved the sound of the horses hooves clippity-clop on the tarmac; however, once the town was behind them the snow lay clear on the road.

'Let's canter,' said Betty. Lilley was first to respond, leading her rider at a steady pace before Bill. Betty was wearing a gaily coloured silk headscarf tied under her chin, the tassels of which fluttered in the breeze. She rose easily to Lilley with a comfortable erect seat. 'She has obviously been taught to ride,' thought Bill to himself, 'which, my lad, is a damned sight more than you have ever been!' But he would not have had it otherwise. His final mastery of Mary and her untimely death had been one of the key points of his life.

Bill had caught up with Betty now. They trotted along together, revelling in the air and the scenery. Her face was flushed and her fair hair, protruding from her scarf, windblown. 'What a place this is,' she said at last. 'Just look at the rays of the sun shining on that side of the mountain. You can almost see the blue of the sky reflected in the snow.'

'Yes,' said Bill. 'I only wish that all this could go on forever!'

'Well,' said Betty more practically, 'let's make the most of it while we have it. Come on, Lilley . . .'

Round a corner another horse appeared, its rider in typically Tyrolean garb, his soft hat and feather matching a tailored riding suit with leather leggings and boots. He carried a whip which seemed more for effect than for purpose. His mount was jet black, with the typical white star on its forehead, its coat sleek and groomed. The harness was also black, with brass buckles that shone in the sunlight. At the horse's heels were two German Shepherd dogs, their ears cocked as though waiting for something to happen. All this Bill observed almost subconsciously. Comparing the other horse with his own mount, he felt something of a 'poor relation'.

As the rider drew nearer, he could see a weatherbeaten face with a bushy greying moustache and somewhat hawkish eyes. As he approached he raised his whip in greeting:

'*Guten Morgen Herr*,' he said to Bill with a slight bow. Turning to Betty, doffing his hat, '*Fräulein*,' he said. '*Von wo reiten Sie, heute?*'

'*Aus Seefeld, mein Herr*,' Bill replied. He was about to return the enquiry, but something made him think better of it.

'*So, gute Reise nach Griesenbach!*' said the horseman. Lifting his whip once more, he continued on his way, calling the dogs. One of the Alsatians took a longing glance at Fritz's heels before obeying his master's summons.

'I wonder who that was,' said Betty as soon as he was out of earshot. 'I thought that he was going to ask us back to his castle or something.'

'He was certainly very courteous,' said Bill. 'But he had a steely look about the eyes. I think he guessed that we were English.'

'Perhaps it was my headscarf. I haven't seen any others like this around here,' mused Betty, tactfully.

They did not reach Griesenbach. After an hour's riding they came upon a small *Holfbräuhaus* which advertised hot meals. 'I'm peckish,' said Bill. 'Let's see what's on offer.' There was a rail for them to tie up the horses before entering a bar, with a roaring log fire and neatly set dining tables. Finding a corner for them to sit, Bill ordered Rheinwein and hot soup with rolls and butter. By the time they had finished, Bill had learnt much more about Betty – of her family and school life, of her work in her father's office, of her friends at home and of her thoughts about their present trip. She spoke with a quiet confidence resulting from a settled and orderly background; she clearly adored her mother and father. How different his own life had been, thought Bill. He wondered what Betty must think of his efforts to establish himself in a foreign land; of the influence that his mother had had upon his attitudes and ambitions . . . and, incidentally, what *were* his ambitions? For now, it was sufficient that he should display his enthusiasm for Life in general.

Outside, Fritz and Lilley were waiting patiently. Bill helped Betty to mount for their return ride to Seefeld. The sun was low in the sky by the time that they reached to town. Bill added an extra mark to the six marks demanded by the horse-keeper.

During the last few days of the holiday everyone was feverishly trying to cram in as many activities as possible; there was an international ski race up on the Schwartzkopf about which there was much excitement – particularly as some of the local lads from Seefeld were competing. They could only marvel at the dedication of the competitors. Back on their own moderate slopes, Hans Wanner got them to do simple jumps. Not very successfully, but at least they could *say* that they had tried! Hans ended by taking Nesta down the slope on his shoulders – to much applause, except from Nesta herself, who said that she was terrified.

The last evening was at the Hohemunde where everyone was fêted by 'the establishment'. Certainly they had been good customers. They raised enough money for four bottles of champagne. Herr Ober then produced another two 'on the house' so they were well primed. Later they monopolised the café, singing 'Auld Lang Syne' in the proper fashion, revolving round the room and raising their arms as they converged into the centre. Beckoning the other guests, soon everyone joined in, many not knowing what it was all about. Later, the manager came to ask what it all meant as he had not seen it before. The hilarity continued until an officer of the *Polizei* arrived to remove them. The poor man ended by joining them for a drink! Finally, better sense prevailing, the officer accompanied them to the chalet, beseeching them to make less noise, whereupon Douglas relieved himself in full public view – somehow, it did not seem to matter.

Next morning their train for Munich left at 8.30 a.m. Over the 7.00 a.m. breakfast Bill and Douglas were still feeling the elation of the evening before and were in hysterics over the antics of the member of the *Polizei*. 'Sober up, you two,' said Irene sternly, 'I shall want some help with the luggage, and mind that *you* are ready.'

And they did. Soon, all the cases were on the way to the station and a final look round found nothing left behind. There was quite a gathering to see them off. Franz the Hausboy and Freda the Hausfrau; the two Totenkopf Guardsmen whom the girls had befriended and even Kurt from the Hohemunde. So it was Auf Weidersehen Seefeld!

The excitement of their journey out to Austria was now replaced by a more calculating appreciation of their surroundings and all that had happened while they had been away. They had all made friendships, many of which would be long lasting. A particular satisfaction to Bill was that the language that he had learnt at Housey, and which had lain dormant for over five years, had become alive once more. But they were already beginning to talk of Home, particularly of the news.

At Munich they caught the train for Cologne, but were sorry to hear that it would be after dark when they reached the Rhine Gorge. However, after leaving Mainz the lights of the boats on the river could be seen in the distance. After a night in Cologne they crossed the border at Aachen without incident, except that Elizabeth had failed to keep some of her bills and was accused of taking German money out of the country. They were glad to reach Dover after a rough crossing and pile into a third-class compartment with plush seats instead of those awful slats. Everyone else was talking English and, well, everything was just *ordinary* again. There was an excellent meal on the train before reaching Victoria. There had been such *esprit* amongst them during the past three weeks that most had a lump in the throat in parting. Douglas had to rejoin 82 Squadron in the morning and would catch a train around midnight for Cranfield. Nesta's parents were waiting for her and for Victor. Both Irene's and Elizabeth's people were waiting for them. Elizabeth had kindly invited Bill back to her home in West London. Before piling into Mr Gorman's car, Bill helped Joan and Betty with their luggage and hailed a taxi for them to Waterloo.

Betty held out her hand. 'It has all been great fun, Bill,' she said. 'You must come and stay with us at Bagshot; I'll be in touch.' Their eyes met and held for a moment in a warming smile.

'I'd like that,' said Bill.

The Gorman household was a luxurious establishment with uniformed servants and double damask napkins. Elizabeth and Bill were expected to eat a second dinner, but neither was prepared to refuse the hospitality. Bill struggled, doing his best, and was glad when the meal was over. He hoped that he was good company, but was dog-tired and thankful to crawl into bed.

Next morning Elizabeth drove Bill with his luggage to the garage where he had parked his car.

They had seen a great deal of each other during the holiday. Bill liked her forthrightness and rather unfeminine competence. As she left, she turned to him with a warm kiss and smile. They both knew that it was a kiss of 'Goodbye'.

5

FOOTLOOSE AND FANCY FREE

What does a young man of twenty-five, with nearly four months' leave in England and quite a decent salary, do with his time?

Such was the problem that Bill found himself facing. He first reaction was, perhaps strangely, one of anti-climax. the last three weeks had been so hectic that he had not thought too much about the rest of his leave. He rang Dolly to say that he was back, to her obvious relief. She had arranged to rent a house at Middleton on Sea on the south coast. They would move from Compton as soon as he got back. Then he rang his firm, Liptons, in the City in response to a request, before he left Ceylon, that he should visit. The Managing Director would see him the next day. Bill rather dreaded the 'interview'. In the meantime he went to the Overseas Club.

Next morning Bill caught the Underground (an experience that still gave him something of a thrill) to Old Street station before walking to the City Road where Offices Lipton Ltd was emblazoned in bold letters on the portico. The receptionist seemed a little surprised when he asked to see Mr Purvis, the Managing Director, and discreetly ushered him into the office of one of the executives, from where, after a few minutes of somewhat desultory conversation, he was ushered into the great man's room. He was sitting in front of a huge picture of Sir Thomas Lipton in yachting attire, a cigar protruding from his fingers, before the wheel of his sailing boat at Cowes.

'So this is Mr Baker from our Pooprassie Group,' said Purvis. 'Take a seat, Mr Baker.'

'Thank you, sir,' said Bill, settling himself into a thickly padded and upholstered armchair. He was offered a cigarette, which he declined.

'Mr Brazier tells me that you play rugby,' said Purvis.

'Yes,' Bill replied, 'I have played both for the Uva and Kandy Provincial teams, and have been in the Up Country Trials.'

'Good,' said Purvis, 'we like our Assistants to take part in sport, and in the social life of Ceylon. It is good both for you and for the image of the company. Tell me about your work on the estate . . .'

They talked for a while. It was obvious to Bill that Purvis knew quite a lot about both Pooprassie and Dambatenne. Purvis intimated that a new tea factory was planned for Pooprassie on a site not far from the Lower Bungalow. This was news to Bill. Finally, he confirmed that Bill's salary had been increased to five hundred rupees a month.

As Bill rose to leave, Purvis smiled at him. 'By the way, are you going to get married while you are on leave?' he said.

Bill flushed at such a personal question. He returned the smile. 'I'm afraid that I cannot say, sir,' he said.

Afterwards, Bill was taken out for lunch 'on the firm' and sat between Snelling the Principal Buyer, and the Managing Director of Maypole Ltd. who was also a guest.

Until then, Bill had always regarded the office wallahs in the firm, both in Colombo and at Home, with a certain amount of suspicion, on the grounds that they might interfere without understanding the nature of things. However, he found Snelling most helpful and approachable. As he left the factory, Bill, in thanking him, said that he hoped one day to be able to show Snelling around his own 'patch'.

It was a cold and blustering March day when Bill set out for Compton House. The little Austin was warm and comfortable and still smelt of new paint. He reminded himself that he would soon have to get the oil changed at the garage. He was greeted at Compton like someone who had returned from the South Pole. Dinner had been laid in the dining-room on the large polished mahogany dining table. Avice had been down to the inn in the village for a bottle of lager. They were 'just dying' to hear all about his adventures. Dolly was a little more reserved. She was just glad that he was back – and in one piece without a broken ankle!

After dinner they sat in the drawing-room, over the customary log fire, drinking coffee from Meissen porcelain cups and choosing bon-bons from a silver entrée dish. Bill found it easier to talk in those circumstances.

The house that Dolly had arranged to rent was ready for them. So, as soon as they could decently leave, Bill packed their luggage into the Austin and kissed Avice and Kathleen goodbye. They had been kindness itself. Dolly had been greatly refreshed by her extended visit and the renewal of such an old friendship.

The house at Middleton was a small, three-bedroomed detached dwelling on a comparatively new housing development. Quite a contrast to Compton House! However, it was well furnished with gas fires in every room. The sea front was about two hundred yards away and there was a tennis club at the end of the road. After unpacking, Bill took Dolly into the village for a meal. Later, after a hunt in the airing cupboard, Bill was glad to find hot water bottles to put in the beds.

Two days later Avice rang. Irene had been trying to contact Bill from London – would he please ring her? It seemed that the Alpine Ski Club had arranged a reunion party the following week at Elizabeth Gorman's house at 8.00 p.m. Irene had asked that Bill should bring any photos he had taken. What a nice idea!

The secretary of the Middleton Tennis Club called upon them and was interested to hear that they were from Ceylon. He introduced himself as John Denning; his wife, who helped with the catering, was Gladys. There were two retired tea planters who were members, but he thought that they had been in India. One could book a court on any day. The Club evening was Saturday for Bar and Bridge.

'*Do* come and join in,' he said on leaving. 'We need some unattached men around the place,' looking at Bill meaningfully.

On hearing about the Ski Club party, Bill had rung Betty Francis to invite himself to tea on that day and to offer Joan and herself a lift up to London. He duly arrived and met both Betty's parents who were most charming. They had obviously heard much about Seefeld. It was nice to put a face to at least one of the participants! The evening was for 'black tie' so they changed after tea. Bill had been invited to stay the night at Bagshot; Betty showed him his room – he surmised that one of the children had been turned out for the occasion!

It was nice to be with the two sisters again; there was an easy flow of conversation as he drove them up to London. Bill took them out to dinner at the restaurant in Kings Street. Remembering Joan's choice of wine on the train in Belgium, Bill ordered a bottle of Graves. He showed them a preview of his photos, causing much amusement.

It was past 8 o'clock by the time they reached the Gormans, where there was a general ovation. All the gang were there, including – surprise, surprise – Noel back from Czechoslovakia. Douglas had got special leave. There was a general exchange of photos and requests for prints to be copied. Both Noel and Victor had films to show – the latter's were in colour. It all created an atmosphere for reminiscing. Nesta asked for a copy of one of the pictures of a Totenkopf guardsman, causing something of a titter. Elizabeth was accompanied during the evening by a tall and rather handsome young man who, it appeared, was a member of the Gorman's firm.

Bill managed to edge Noel into a corner. 'I have been thinking a great deal about our conversation before you left the party,' he said. 'Did you have any trouble getting into the Sudeten, and how did you find your mother?' he asked.

Noel hesitated before answering. 'Well, I was asked, closely, why I wanted to go, and had to produce a letter from my mother asking me to come – it seems extraordinary that I had to prove the reasons for such a visit! They also checked me out. In the country, the Czechs seem to expect Hitler to invade. There is even some opinion that their economy would improve under German rule, if Hitler were to be allowed to remain in possession, that is. I doubt whether the Czechs will resist Hitler. What they all wanted to know from me was whether an invasion would bring Britain into a war, and that of course I could not answer. I hated leaving my mother there, but she will never leave her homeland.'

'That must have been horrible for you,' said Bill. 'The future just seems to be in the lap of the gods.'

There was laughter coming from the other end of the room. Betty was describing her experiences in thwarting the attention of the Woolly Bear. Bill also heard his name mentioned in connection with a certain *Weinstube* in Munich. He left Noel to deal with the storyteller. Then there was rum punch and sandwiches.

Bill gave Irene a lift home on the way back to Bagshot. Then Irene gave them minute instructions on how to get onto the main road, and Betty was quite sure that she knew the way. Happy to leave her to do the navigating, Bill was chatting away merrily recording his impression of the evening.

'Turn right here,' said Betty with conviction. 'And again here; this will take us through Richmond Park.' Whereupon the road came to a dead end.

'Now where?' said Bill, kindly.

'I'm afraid I have no idea,' said Betty, crestfallen. A street lamp cast a clouded beam across her face; poor girl, she looked so ashamed that Bill burst out laughing. Soon, all three were convulsed. Bill put his hand over to touch hers. 'That's the best laugh I've had all evening.'

Eventually they found the Park and the main road. It was 1 a.m. by the time they returned to a darkened house. Betty showed Bill to his room. 'Breakfast is a help-yourself affair,' she said. 'I'll give you a shout when the bathroom is free. Both Joan and I have to be at the office quite early.'

'Thanks,' he smiled at her. 'It's been a grand evening, Betty, I have enjoyed it all.'

'So have I,' she responded as their eyes met. 'It's been fun. Thanks for driving us up to London.' Instinctively she threw her head back as she turned away, causing the fringe of her fair hair to catch the dim landing light. As he lay in bed, Bill's mind ran through the day's events. There was a warmth within him that came not only from the bedclothes.

Breakfast was, indeed, a family gathering. There was a younger sister, two boys and little Ruth who was only four and who was everyone's pet. She insisted on sitting upon Bill's lap and sharing his cereals. She offered him in return her glass of milk, until Bill pointed out how important it was that she should drink it in order that she should grow up like her big sister. As she followed Joan down to Bagshot, Betty held out her hand. 'Goodbye, Bill,' she said. 'We must meet again; I'll be in touch.'

Taking his leave soon afterwards, Bill thanked Mrs Francis. She was a charming person – obviously proud of her family and of her home. 'You all seem to have had a wonderful time in Austria,' she said. 'Betty seems to have talked about little else since she has returned.'

'I hope that we shall see you again,' said Mrs Francis as Bill waved goodbye.

As always, Dolly flung out her arms to embrace him on his return. 'Lunch is ready,' she said. 'Come and tell me all about your trip up to London.'

When Bill had finished a somewhat colourful account of the reunion and of the Francis family at breakfast that morning, Dolly intervened. 'We have been asked by the Dennings to make a four at Bridge at the Club this evening – I've accepted on you behalf: I hope that that's all right.'

'Ah well,' he replied, 'I suppose that we must start sometime!'

'It's 3d. per hundred for new members and 6d. for 'regulars',' she continued.

The time arrived for Bill to report at Winchester for his attachment to the Rifle Brigade. Entering the Adjutant's office, he signed on with the minimum of formality as a temporary sergeant. A regular sergeant, James Condy, who was about his own age, and whom Bill guessed was earmarked for an Officer's Training Corps, was assigned as his instructor. At the Quartermaster's store Bill drew fatigues, boots and tie-on stripes. He was allowed to wear his Ceylon Planters' Corps cap.

The battalion was at full fighting strength, both in manpower and equipment. There was an air of promptness and precision about everything in which Bill soon found himself involved and carried along. He was to attend Daily Orders and was issued with maps and a compass. Their platoon commander was 2nd Lieutenant John Rolt, who in his spare time was a racing driver who performed at Brooklands. Rolt eyed the Colonial Sergeant with some condescension until they found a mutual friend – a broker with McKinnon Mackenzie in Colombo.

That evening Bill was in the Sergeants' Mess shepherded by James. This was something of an experience; he tried hard to understand what was being said. One sergeant in particular, whom everyone called 'Gutsie' and who came from the East End of London, used expressions that Bill, for all his experience, had never heard! Gutsie's language left little to the imagination, but his enthusiasm for the brigade, and those in it, was overwhelming. Bill soon felt that, if ever he was in a tight corner, Gutsie was the one he would like to have with him. The other members of the mess, though not so colourful, were of the calibre that makes such a fine battalion. Bill chose to listen rather than to expound. James introduced each as the opportunity arose. The signals sergeant John Fletcher was interesting. The CPRC was particularly short of signal equipment; John promised to take Bill round his store.

There was a sudden silence and everyone rose to their feet. The Regimental Sergeant Major, complete with black belt and red sash, had entered. As the RSM came in his direction James said: 'Good evening, sir.'

'Good evening, Sergeant Condy,' was the reply.

'What's the news, sir?' asked James.

'Not good, I'm afraid. The Boche is moving two more Panzer brigades over to the Czech border.'

'Why should Hitler do that, sir?' asked Fletcher.

'They say that he thinks Britain won't come in, if he goes into that country. He's taking the risk.'

'Then we shall be able to shoot him up the backside when we do,' said Gutsie with evident relish.

'Sergeant Baker has been over there recently,' said James, by way of introduction.

'Have you now, Sergeant?' replied the RSM as he moved towards the bar. James looked a little embarrassed, but Bill took the subject as being closed.

Next day Bill joined the battalion parade. Feeling a little self-conscious, he fell in, with James, at the rear of 'A' Company for the CO's inspection. As the CO passed down the files of his men, Bill caught a glimpse of his face; he was not the tall, stalwart military figure that one would expect, but of medium height with slightly greying hair. His eyes were kindly, though penetrating. He wore the DSO above his First World War medals. Such was the man who would lead the battalion into war once more, if necessary.

The CO looked at Bill as he passed, but did not speak to him.

The days passed quickly; there were two sessions on the ranges, one with rifles and one with automatic weapons. James handed Bill a Lee Enfield. 'Have a go,' he said, 'you're on No. 3 target.'

Bill had been waiting for this.

'Two sighters and ten rounds application: Fire!' continued James.

After the first two sighters, James looked approving. 'Well done! We will not have to teach you how to shoot.' Following through, Bill scored 36 out of 40 on the application. 'You will qualify!' remarked James with a grin.

Bill did not think it necessary to say that he had already qualified for the year on the ranges at Diyatalawa. 'Good,' he said, 'we'll have a drink on that in the bar this evening.'

Later, during a practice using automatic weapons, Bill took a section of men marking in the butts. He was glad to be given a job and it was quite an experience to have the rattle of Bren gun fire as it passed through the screen, a foot or so above his head.

After returning from a field exercise in which his company was in support of an attack by Bren gun carriers and a detachment of anti-tank guns, Bill was handed a message. Lieutenant Lewis, Battalion Intelligence Officer, would like to see him at his office at 1700 hours. What was this about, Bill wondered.

On entering, Bill was greeted by Lewis with a smile. 'I'm Tom,' he said, 'and I gather that you are Bill, and that you are a tea planter on leave from Ceylon.'

'You are well briefed,' said Bill. 'It is nice to be here; my unit is affiliated to the Rifle Brigade in some way.'

'I also gather that you have just returned from Europe – was it Germany or Austria?'

'Well, both really, but mainly Austria, and that was for only three weeks.'

Lewis continued: 'Of course we get our official intelligence reports of what is happening over there, but if you had contact with the local people I would be interested in any impressions that you got.'

They chatted for some time – Bill described their visit to Munich and the Fasching Festival; what he had heard about 'Kristallnacht' and what he had been told about the situation in Czechoslovakia, all of which seemed to be of interest to Tom.

When they had finished talking about Bill's visit to Austria, Tom changed the subject. 'I am Vice-President of the Officer's Mess' he said. 'The CO has asked me to invite you to join us for dinner on Saturday night. We shall be in Mess dress – I believe that you have a dinner jacket?'

'Yes, I have, and I shall be delighted to join you,' replied Bill.

'Good; then perhaps we should meet here at 19.30 hrs. I will arrange for you to sit next to me and I will introduce you.'

Bill saluted as he left the IO's office – so he was to be a civvy on Saturday evening! He would look forward to it.

The dinner night arrived. Tom Lewis was waiting for him when Bill arrived at his office, and they went over to the Officer's Mess. As he was ushered into the reception hall, Bill was overawed by the vivid paintings of past regimental campaigns through its long and gallant history. There were scenes of the most bloody conflict. Busts and statues of earlier commanders, in appropriate uniforms of the time, occupied alcoves along the walls with flags and battle honours draped above them. There were also cups and trophies of peacetime excellence displayed and illuminated in glass cases. The CO stood at the end of a long carpet bearing the regimental crest, where he welcomed his guests. He asked Bill about the CPRC and enquired the name of his CO.

The dining table, which ran the length of the room, was set with six huge silver candlesticks. The light from the candles was reflected in the polished mahogany table. Also on the table were three huge silver rosebowls filled with spring flowers. (Tom apologised that, at that time of year, there were no roses!) The silver flatware glistened in the flickering light, but perhaps the most remarkable of all were the starlike reflections in the crystal goblets set to each place. As they took their allotted positions, the officers looked resplendent, if a little sombre, in their dress uniforms. Everyone remained standing for the CO to take his place, whereupon a host of batmen appeared.

As Vice-President, Tom was on Bill's right at the lower end of the table facing the CO. On Bill's left was a young 2nd Lieutenant recently commissioned. Rolt was further up in seniority on the opposite side of the table. Bill enjoyed the meal – his companions were welcoming and interesting. As the dinner drew to a close the CO rose, goblet in hand. 'Mr Vice; the King.'

Tom rose solemnly. 'Gentlemen, the King.'

Then there were the cigars and coffee accompanied by the most copious brandy goblets that Bill had ever seen!

Bill crept back into the Sergeants' Mess by the back door. He did not wish to be seen in his black tie, nor particularly, to be engaged in further conversation.

Church parade on Sunday was held on the parade ground. The Chaplin, resplendent in his vestments and supported by three Servers, appeared upon a dais with the regimental band before him. Bill found the music, and the lusty singing of multiple male voices all around him, to be truly uplifting. There were the old hymns that he loved, and, of course, Onward Christian Soldiers. Almost ashamed, he had to hold back a catch in his throat. Then the mood of the service changed with an abridged form of the Communion. Each man filed past the dais to receive the Sacraments, while the band played muted music.

The Chaplin entered the Sergeants' Mess afterwards; Bill had a quick word with him.

Bill had arranged to leave Winchester after church parade and had already handed in the clothing that he had drawn. Gathering at the bar of the Mess, he stood a farewell round promising to send James Condy a box of tea from Pooprassie. John Fletcher was there and also, of course, Gutsie, from whom Bill got a particularly knowing wink, though, in reflecting afterwards, Bill wasn't quite sure why!

On the way out, he tried the Adjutant's office, but it was locked. No matter, he thought; in any case he would have to write a note of thanks to him and also to the Mess Vice-President. He supposed that he would also have to write some sort of report for his own Adjutant – he had better make some notes!

After his exertions in the previous week. Bill was content to relax in the little house at Middleton. He enjoyed walking along the seashore and watching the waves roll in from the Channel. There was still a bite in the air – he had to keep moving. He missed the company of Boots at his heels. The wooden groynes jutting out into the sea to prevent the shingle from moving reminded him of his childhood days on the beach at Bognor; now, the rollers appeared to beckon as if in challenge. He gave a little shudder. Not today, (or tonight) Josephine!

Next day Dolly announced that she had invited her niece, Ann Russwurm, who was unmarried and about Bill's age, to stay for a week. They picked her up at the station and drove back to the house for lunch. She was demure and thoughtful, helping Dolly with the crockery and drying up afterwards. Later, Bill drove them to Chichester to see the cathedral.

Reverend enough as they entered, inside, some of Ann's coquettishness returned with some most 'enlightened' comments on the attitudes and attire of some of the past dignitaries whose effigies were amply provided. 'Ssh!' said Bill. 'They might hear you!' Inwardly, he was inclined to agree with her. They had tea near the cross. Dolly was pleased to have her niece for company and Bill found her simple and uninhibited presence unusually refreshing. In the evening they played three-handed whist.

The next day, Dolly had some shopping to do, so Bill drove Ann to the valley of the river Arun, near Amberley. It was a lovely sunny day with the signs of spring around them.

'Look,' said Ann. 'There's a moorhen with her chicks.' As they watched, the clutch swam out to mid-river, all that is save one, that had become caught in the reeds and was chirping madly. Ann bent down to free the little thing, nearly loosing her balance. Bill caught her hand to pull her back. 'Thanks, Bill!' she laughed, 'that was a near thing.' Further along a pair of swans were watching them. The pen was brooding peacefully on her nest on an island in the river, the cob swimming gracefully in an attitude of protection.

'Those birds mate for life,' mused Bill. 'I wonder whether that is instinct or intelligence?'

'If its the latter, they must be able to talk to each other,' laughed Ann. 'Perhaps the pen says: "Cob, give me some more feathers for the nest!"'

The next day Bill took Dolly and Ann for a drive through the Sussex countryside. Taking an Ordnance Survey map from the pocket of the car he lent over to Ann to show her the route. 'We'll go through the gap in the South Downs at Upwaltham,' he said. 'At the foot of Duncton Hill we will turn east, and follow the minor roads that run along the northern face of the Downs until we reach Washington. Then, if there is time, we could climb up to Chantonbury Ring to see the view.'

Ann looked at the map blankly. 'What are those red lines on it?' she asked.

Bill's heart sank. 'Those, my dear, are the roads,' he explained, gently retrieving the map as he spoke. He would have to be his own map reader!

As they passed through the village of Bury, they came upon a particularly fine example of the timber-framed building and stopped to admire it. Built upon massive-gauged stone foundations, the lintels above the ground-floor windows were carved with fleur-de-lys alternating with the English rose. The joists of the first floor were extended over the lower wall to give a pleasing proportion. These were trimmed, and also carried decoration.

'What a lovely place to live in!' exclaimed Ann.

'Yes,' said Bill. 'They say that many of these houses were built with ship's timbers salvaged from as long ago as the Armada. If they could talk, they cold tell a story or two.'

Emerging from the woodland, the chalk downs lay open before them. What a contrast! South Down ewes were grazing peacefully with their lambs. Many had twins, either jostling each other for their mother's milk. or scampering carelessly over the tight downland sward.

Reaching the end of the road, they walked the last half mile or so to the Clump. The breeze was noticeably stronger here causing the beech leaves, and even the branches, to respond with rhythmic movement. When they reached the crest they stood beneath the clump of beech to drink in their surroundings.

'This air is like a tonic,' said Dolly breathing deeply.

'And just look at that view!' said Ann.

Bill had been doing some homework before they left. 'The guidebook says that this clump was planted about 1760 to enclose an Iron Age hill fort. It also says that there was a Roman temple on the centre of the site.'

'Lovely as it is to visit, I shouldn't like to live up here,' said Ann with a slight shiver.

The week soon passed, with a trip to Brighton and two evenings at The Club. They spent the last morning chatting before Ann went to pack her things. Dolly came with them for the drive to the station at Worthing. Bill kissed Ann as she boarded the train, and she was profuse in her thanks to Dolly. Turning finally to Bill, 'We must meet again before you return to Ceylon,' she said as the train pulled out.

Bill had been planning, during his leave, to take Dolly for a tour of the West Country. They would stay a few days with his late father's brother, Tun Baker, and his wife Hilda, call upon the Gormans at Marlborough and then explore Devon and Cornwall. Dolly was looking forward to this final jaunt with her son. After Bill had returned to Ceylon, she proposed to return to Canada to live with her sister Kitty at Stewart, British Columbia. Bill and Dolly would be at opposite ends of the world. Fate would then have to take care of itself.

On the way to his uncle's cottage at Harwell, Berkshire, they passed close to Bagshot, so Bill called upon the Francis household to introduce Dolly to Mrs Francis, who was delighted to see them. The two were glad to meet after all that they had heard about the Austrian adventure. Bill was content to let them talk. While they were drinking coffee, Joan came in to take her mother to the dentist, and suggested that they should walk down to the solicitor's office in the village where Betty worked for her father.

A receptionist ushered them in. Betty was on the telephone at her desk, but there was instant recognition and pleasure on her face as she signalled apology and guided them to two antique walnut chairs in front of her. It was the first time that Bill had seen her occupied thus – it gave him a moment of contemplation. The offices were in part of the terrace of seventeenth-century cottages with dark oak beams and small leaded light windows which faced a back yard. Floorboards seemed to give as one moved across the carpet. On Betty's desk was a green-shaded reading-lamp which cast a halo of light over her face and her fair hair neatly brushed back over her forehead. She spoke with an air of quiet confidence, which Bill, had he been on the other end of the phone, would have found most reassuring. She was making notes as the conversation progressed. Her coat and scarf were hanging neatly on the door, but Bill spied a basket of groceries in the corner behind her desk.

Finally, putting down the telephone, Betty came round, smiling, to greet them, 'Oh, what a lovely surprise!' she said.

'We were just passing on our way,' said Dolly. 'It seemed an opportunity not to be missed.'

'I'm so glad you did,' replied Betty, 'I was just going to ring you, Bill, to suggest that you joined us for a small tennis party next weekend. We are playing at the Pantiles Club – close to us.'

Sadly, Bill had to decline as they would still be away by then. Instead, they arranged that he should stay the following weekend when they could play tennis in the afternoon and dance in the evening. He would look forward to that. The telephone rang again – Betty blew them a kiss as they left, which Bill returned.

After lunch in Reading, they followed the Thames valley through Pangbourne, where they stopped for a while to watch the water cascading over a weir by one of the locks. As they walked along the bank towards the lock they were hailed by the lock keeper as he closed the lock behind a boat on its passage upstream. Quite a pleasant occupation for a short time, thought Bill!

As they ascended onto the Berkshire downs they admired the open stretch of country before them with its tightly grazed turf and rolling landscape. Sheep and lambs were everywhere, still unclipped; the ewes were heavy with wool. Through Blewbury with its neat white stucco and timber-framed houses bordering the village street, to Harwell, where they found Tun's cottage without difficulty.

Tea had been laid in a patio bordering his bowling lawn. Tun was captain of the village team and spent many of his summer evenings either maintaining the grass or entertaining players. Everything was of genteel simplicity. A white crochet and linen tablecloth had been set upon a plain round garden table. Hilda, a petite and finely drawn figure, with jet black sparkling eyes below a black fringe of hair protruding from the plaits surrounding her head, poured tea from an ornately decorated silver teapot, replenished from a stainless steel kettle, cradled above a methylated spirit lamp. They ate sandwiches of crustless brown bread flavoured with Gentleman's Relish, followed by fruit cake.

There was not enough room for them all at the round table, so they sat around the patio in deck chairs. Val, their sixteen-year-old son, handed the plates around, leaving them to balance the sandwiches in the saucers to their teacups.

The two sisters-in-law had plenty to talk about so Tun, Bill and Val had a game of bowls, before Val, who was returning to Radley College that evening after the spring holidays, had to leave. As one would expect, Tun had an exasperating ability to land his woods within inches of the jack. Bill persevered before finally winning an 'end'. A call from Hilda to Val to finish his packing left Bill and Tun to roll the lawn after the game. They had exchanged cards at Christmas, but it was the first opportunity for a chat since Tun left Maresfield some ten years previously. A taxi took Val and his luggage. (Rather more than he had been allowed at Christ's Hospital, thought Bill.)

As he left next morning, Bill felt that he had a new perception of his uncle. Now that he had grown up, he supposed. Both Tun and Hilda had been friends, rather than relations. When would he see them again, he wondered.

Out along the Great West Road through beautiful countryside. The leaves were bursting with subtle shades of green and yellow against a cloudless blue sky. The primroses were going over now, but the woodlands were clothed with bluebells and the fields with buttercups and moon daisies.

The Gormans were most welcoming. A Wing-Commander and Mrs Davey had also come to tea and there was a jolly party. On hearing that Bill was going back via Marseilles after his leave, she assumed that he was taking the car overland and asked for a lift! She was quite attractive and Bill would have been glad to oblige, but had to explain that the car was being shipped from England. Gordon Davey remained quite impassive during the discussion! Diane came in after tea looking much the same as Bill had last seen her at Bandara Eliya. She greeted Bill with a wink. He surmised that her mother had never been quite sure what had happened in the guano cave, and Diane had been quite content to leave a little mystery surrounding the event. She disappeared to do some preparation for the following day.

After dinner, Mr and Mrs Gorman partnered Bill and Diane at Bridge. Diane was quite a good player for her age and in the end she and Bill had the satisfaction of taking three hundred points off her parents, amid much jollity. Dolly had already retired when he went to bed in an attic spare room. As he climbed into it Bill found that it had been stuffed with old coat brushes – the first apple pie he had been served since his school days! He resisted an immediate desire for retribution – they were really too old for that sort of thing. Diane had already left for Western when he came down to breakfast – probably just as well!

After leaving the Gormans, Bill and Dolly continued their tour of the West Country. They made an odd couple – this sixty-year-old widow and her twenty-five-year-old son, though Bill did not give much thought to the matter. Typical of youth, he felt no embarrassment, nor, come to that, had he much regard for the future. He and his mother had been very close and used to each other's company. Now, he was glad to give her pleasure before the end of his leave. Besides, it would not have been much fun on his own.

For Dolly it was different. In less than two months she would be parting from her son for a second time. It would not be right for her to return to Ceylon to live with him again, even if he did not marry; nor would she wish to, for it had been rather a purposeless existence for her. Yet he was all that she had in the world – the thought of parting tormented her. Furthermore, there was the threat of war. Had she not suffered enough as the result of the last? But for now she would make the very most of the present, and enjoyed every moment.

Bill remembered the North Devon coast from the days when, as a schoolboy, he had cycled to Saunton. Now, he made for Minehead and the coast road to Lynton. Out of Porlock, the little Austin was boiling by the time they reached to top of Porlock Hill – and he had made it on a push-bike! So much for pedal power.

At Lynton they had no difficulty in finding a cottage, covered with sweet-scented viburnum, offering bed and breakfast. There was time to visit the Valley of the Rocks before dinner at a hotel in the town. Bill climbed up high onto one of the rocks and imagined that it was the one that he and a friend had scaled some ten years previously. A rugged place, oblivious to the passage of time.

A signpost to Clovelly provided a welcome break and some exercise. Leaving the car above the village, they followed the steep, narrow and cobbled street down towards the sea, passing the lines of shops and cottages as they went. Donkeys driven by small boys with sticks, which they had no inhibitions about using, trudged up the street laden with panniers of fish. Down at the harbour it was peaceful enough. Bill and Dolly sat on the quay watching the activities on the boats below them.

'Time to walk up that hill again,' said Bill eventually.

'Yes,' replied Dolly, resignedly.

Heading southwards, and into Cornwall, the Ordnance Survey map that Tun had lent them helped to discover many of the remote and beautiful parts of this coastline, often with surprises at the end of the byway and a welcome from the locals. With time as their own, they could meander as they pleased and drink in the beauty of the scenery.

Bude, Newquay, St Ives and Land's End; these had all been just familiar names which Bill was glad to have translated into actual places that he could picture, helped by his Brownie Box camera. As they stood on the rocks at Land's End, Bill felt something of the stark emptiness of the surroundings. 'Next stop America,' he said, somewhat thoughtlessly. 'And Canada,' Dolly was about to add, but decided to hold her peace.

By now, Bill, while not wishing to hurt Dolly's feelings, was becoming anxious to return to Middleton. He wanted to company of other young people, and wondered whether Betty had tried to contact him about their impending weekend. Besides, it would be fun to play tennis at the Club again with some of the friends that he had made there.

As they turned into Middleton Bill felt excited – would there be a letter from Betty about their weekend? Dolly fumbled with the key from her bag; there was a pile of letters on the mat, mostly bills. At the bottom was one from Bagshot in Betty's fair round hand. It was the first he had had from her. Would Bill like to come up for lunch on Saturday; they could play tennis in the afternoon and have a swim afterwards? There would be a dance at the Pantiles Club in the evening. She hoped that he would stay on for Sunday lunch.

He rang her at once – at the office. Yes, he would, very much. He would bring his dinner jacket for the dance. Betty's 'official' voice mellowed as she spoke; she was looking forward to their meeting too, and also wanted the hear about his trip with Dolly.

Next day, Bill renewed his acquaintance with the Club. It was a sunny May day with everyone sitting on the balcony watching the tennis on the first court. There was a foursome that was just finishing. John Taylor and his wife had the next booking and suggested that Bill should make up a fourth with Jan Collins, a rather bubbly girl whom he had met before. 'Which side do you prefer?' said Bill, as they walked onto the court. Perhaps Bill had misjudged her – she had a strong backhand and quite a determined service. The Taylors won the set, but only after five games all.

They sat outside after the game drinking lime and soda. After a while Jan turned to Bill. 'We have arranged a boating party on the Arun next Wednesday,' she said. 'The idea is that we should take two punts and work our way upstream for a picnic lunch. There are five of us – we need another man to help with the paddling!'

'It's very honest of you to put it that way,' laughed Bill. 'Do I take it that you have four girls and one man, and that you need another for the second boat?'

'Yes,' replied Jan. 'But the girls will help!'

'OK, I'll provide beer and cider – you bring the sandwiches. Do I also assume that you would like lifts into Arundel?'

'Right again; we'll meet here at about half eleven.' By now Bill was beginning to understand why he had been asked, but then, why shouldn't he help?

But there were more important things to think about before the boating party.

On Saturday morning Bill set out on his visit to the Francis family at Bagshot. Dolly had helped him pack his tennis clothes and dinner jacket. (In fact, she had done most of the packing, but Bill could hardly be expected to admit as much.) He had armed himself with a large box of chocolates for Mrs Francis.

It was just noon when he swung into the steep drive leading to Park Hill. Betty answered the door and had obviously been expecting him. She was wearing an attractively coloured summer dress with white sandals; she smiled to greet him. 'Hello, Bill,' she said, 'how nice to see you,' holding out her hand. Their eyes met – he had been looking forward to this.

At lunch Mr Francis sat at the far end of the table with Joan on his right. At the other end, Betty and Bill were each side of Mrs Francis with the younger children in the middle. Bill noticed that having prepared the meal, she sat down and was waited upon by the children. There were no inhibitions – everyone chatted away merrily; it was up to Bill to join in if he could!

Lunch over, Betty was excused the washing-up and took Bill out onto the terrace for coffee. As she was pouring there was a knock at the front door. 'Excuse me a moment,' she said. Returning, she held out a telegram, looking a little apprehensive. 'It's for you, Bill.' He tore it open. It read –

'Evening dress collar left behind – buy one. Mother.'

They both burst out laughing.

'I was going to show you your room. When you have unpacked, we'll walk down to the village. I'm sure that you can get a collar there. I think one of Daddy's would be too big,' she said smiling.

Betty was already in the hall when he came down after changing for tennis. She cut a trim figure in a white pleated skirt with a white blouse and cardigan. She was carrying a string bag of balls and her racquet. Joan joined them soon afterwards – they would meet John at the Pantiles.

The tennis was a none-too-serious affair, with plenty of chatter between games and discussion about what the score should be. It was very lighthearted and immensely enjoyable. Betty was obviously a talented player and there were periods of quite good tennis. Bill found himself being acutely aware of her presence on the court, and sensed a need for he himself to play well. He hoped that this was not a desire to impress!

Joan was not coming to the dance, so Betty had arranged for a school friend and her partner to join them for dinner at the Club. Complete with his new dinner jacket dress-shirt collar, which they had successfully bought in the village, he met Betty on the stairs at Park Hill. She looked absolutely stunning in a plain, full-length and tightly fitting silken dress, grey-green in colour, and with a coloured shawl thrown over her bare shoulders.

'Gosh, Betty, you look marvellous,' he said simply.

'Thanks,' she said. 'We had better go, otherwise the others will be there first.'

On the way, she explained that Jane had been at Manor House School, Godalming, with her when she was a boarder there; she hadn't met her partner.

Jane and John were a delightful couple and the wine flowed. Betty was in her element; Bill saw in her new flashes of merriment and *joie de vivre* that he had not noticed before. The courses of the meal came and went without anyone paying much attention to what was being served. People began to dance.

Bill politely asked Jane for the first dance. Jane had many stories of their times at Manor House which Bill found slightly embarrassing. For the first time Bill learnt that Betty was only nineteen – he had thought of her as at least a year older, she seemed so competent at everything, but that did not seem to matter. She had kept very quiet about her age herself!

Apart from a few Paul Jones, the partners kept to themselves. Bill was drinking in the music, the place and the feel of the girl in his arms. Jane and John excused themselves later but the evening continued to slip away. All too soon the band was playing the last waltz – The Skater's Waltz which was becoming Bill and Betty's 'own tune'.

They walked back to Park Hill holding hands. The moon was in the last quarter – the stars shone brightly above its waning light. 'There's the Plough,' said Bill, 'If you follow the pointers, you can see the North Pole.' Their heads touched as they gazed at the heavens.

'And if you look to the right you can see the Little Bear,' Betty responded adroitly. There was something funny about this reply; they both laughed, Bill looked at her – No, he had no right, he thought. They moved apart – the moment was lost.

'Goodnight, Bill,' she said on the landing. 'It's been a lovely evening.'

'I've enjoyed it more than I can say,' he replied.

He was awakened by Little Ruth jumping up and down on his bed, followed by Betty, in a trim, brightly coloured dressing gown, with a cup of tea.

After lunch, Betty took him for a long walk along the byways and footpaths that she knew so well. Sometimes they strayed onto private land. 'It does not matter,' she would say, with a note of pride in her voice, 'this land belongs to Old So-and-So who is a client of Daddy's.'

They came to a wooden bridge over a little stream. Plucking two twigs from a nearby bush, 'Let's play pooh-sticks,' she said lightheartedly. They lent over the side rails upstream, throwing the sticks into the water. Then, running to the other rails, 'There's mine!' said Betty, laughingly. 'It's beaten yours!'

'It's got stuck in some weed,' replied Bill grudgingly. 'Let's try again . . .'

Bill had told Dolly that he would be back for supper. Betty came up to his room to help with his suitcase, sensing correctly, from his mother's telegram, that he was not exactly accomplished in that respect. That finished, he turned to her.

'Betty, could you come down to Middleton for the weekend? – it would be lovely to have you, and Mother is so anxious that you should.'

She thought for a moment. 'I'd love to,' she said after a pause, 'I was just wondering how I would get down.'

'I would fetch you and bring you back,' said Bill.

'No, wait a moment. Joan could drive me to Woking station and I could come by train to Portsmouth. Perhaps you could pick me up from there?'

And so it was to be – the week after next. There were goodbyes and thanks for Mrs Francis. He held Betty's hand, releasing it reluctantly before departing.

Dolly was glad to see him back and supper was ready. 'By the way, Jan Collins rang while you were away to remind you about the boating party on Wednesday. She seemed rather interested in what you were doing up at Bagshot . . .'

'Was she, by jove?' replied Bill. 'By the way, Mum, I have asked Betty to stay for the weekend after next. She will come down after the office closes on Friday night and stay until after tea on Sunday. We will pick her up from Portsmouth station. I hope that that is all right.'

If Dolly felt any surprise, she did not show it. 'Of course; she is such a nice girl, but I shall have some planning to do.'

Bill had some other matters to attend to before the boating party. He was due to leave England, on his return to Ceylon, on 11 June. He planned to stay the night in Paris before catching the express for Marseilles, on the 12th, where he would embark. The contract for the exchange of his car for the new one in England had been arranged in Colombo and included the shipping of it back to Ceylon.

The garage in London wanted his car for shipping by 5 June. All being well, it would be on the quay at Colombo when he arrived. The garage would arrange to hire him another car for the period 5–11 June.

That arranged, Bill felt free to face the boating party. With the beer and cider packed away in the boot, he picked up Jan Collins and two other girls, Sarah and Pru, before meeting the future occupants of the other punt on the way to Arundel. It was obvious that it was to be a boisterous affair; Bill had managed his white flannels and sports jacket, but the other man sported a 'boater' as well. The girls were all flimsily dressed in white with large straw hats and white gym shoes. Jan produced a wicker hamper which took up the whole of the boot.

Aboard the punts things began quietly enough, with Bill paddling from the rear. He would have been content to be the single paddle, steering by pushing the blade outwards at the end of each stroke to maintain direction. However, Jan insisted in grabbing a paddle for the other side and in using it with youthful exuberance regardless of direction. 'Hold hard, Jan!' shouted Bill, but it was too late. Poor Sarah, who was in the bow of the boat, disappeared amid screams under the boughs of an overhanging willow. 'Are you all right?' asked everyone, as the boat backed out. Sarah was retrieving her straw hat from the water. She gave a non-committal reply and a slightly accusing glance at Bill. There's no justice in this world, thought Bill.

The afternoon passed without anyone noticing the time. It was nearly supper time when Bill arrived home. In retrospect, although he had been rather dreading the day, he enjoyed it; they were a cheery bunch. He would have to tell Jan so when he next saw her at the Club.

Bill was in eager anticipation of Betty's visit at the weekend. The more he thought about her, the more he felt that she was a kindred spirit – one with whom one could be completely relaxed and just be oneself. Perhaps it was that with her family background, she had no need to put on the airs and graces that others seemed to acquire. Besides, she was so thoughtful of others, and they enjoyed each other's company. It bothered him that he would soon be leaving her behind.

Dolly was imperturbable about the visit: there would be supper on Friday, breakfast on Saturday and Sunday. What would they be doing on Saturday evening? For once, Bill was undecided. 'Could we wait and see, Mum?'

'Of course,' she replied, making a mental note that they might be in for supper.

Bill met Betty at Portsmouth station as planned. It was a long train, but he picked her out immediately. They met with obvious pleasure, shaking hands. Then he put his arm through hers to lead her to the car. She was dressed in a thin navy blue straight cut and waisted coat, above a simple flowered summer dress with a white silk scarf thrown around her shoulders. Bill looked at her approvingly, but did not comment.

They chattered away on the return to Chichester – even though they had only seen each other a week previously.

'I'm learning to drive on Daddy's car,' she announced.

'Then you must allow me to be your instructor in this one,' said Bill. Stopping at a roadside garage, he bought some 'L' plates. 'There,' he said, 'tomorrow we will find a quiet road somewhere for you to practise.'

Dolly had taken great care over the meal. After avocado 'starter', there was roast chicken with bread sauce, green peas and new potatoes followed by fresh fruit salad and ice cream. Bill had opened a bottle of Liebfraumilch for the occasion.

Dolly had noticed the elation in Bill's eye when speaking of this fair-haired young girl; now, seeing them together her heart warmed to Betty. She seemed so dignified, yet easy and capable. After the meal she insisted on joining Bill for the washing-up, saying that this she always did at home.

Bill suggested that they make the most of the fading sunlight by walking along the beach and to watch the sunset. As they reached the seashore the sun had almost submerged below the sea. 'See if you can see a flash of light when it goes!' said Bill. 'There now, until tomorrow.'

Arm-in-arm they strolled along the shore, the gravel crunching beneath their feet. Their experiences in Austria were still much in mind. 'I heard last week that Nesta has become engaged to the son of their local GP,' said Betty. 'The fiance is a professional ballet dancer – the engagement has caused quite a stir.'

'Well,' said Bill, 'Nesta was always rather abstractly and artistically inclined. I hope that they are happy.' After a pause – 'Any news of Elizabeth?' he enquired.

'No, none,' replied Betty. 'You rather liked her, didn't you?' she added, a little provocatively.

He held her arm a little closer. 'Did I give you that impression? We have not seen each other since we returned. She had many other interests in England.'

There was a chill in the air now that the sun had gone. They walked a little faster.

'I did enjoy that evening we had at the Pantiles,' said Bill, 'and often think about it. Do you go there often?'

'Not very often to dance,' she replied. 'But we play tennis fairly regularly in the summer.'

Dolly had retired when they returned. Bill led Betty up to her room. 'I hope you will find everything you want. I'll bring you a cup of tea in the morning.'

'Goodnight Bill,' she said.

They were in no hurry in the morning, lingering over breakfast, while Dolly filled up the coffee cups. Eventually Bill suggested that he should take Betty for a drive and give her a driving lesson. After fixing the 'L' plates, they set off for Climping, where Bill knew of a back road up to Ford and Eartham, where there would be little traffic. He found an open space for parking.

'How far have you got, Betty?' he enquired. 'Have you been driving yourself yet?'

They sat in the car for a while talking about how the car works. Bill had brought a pencil and paper, using them to sketch the various functions of the engine, the gearbox and the drive to the wheels. Then they got out for Bill to show her the engine, which, in the Austin 7 was notoriously simple. Back in the car, he demonstrated starting, stopping and changing gear, each of which Betty did a number of times in the driver's seat.

It was now a bright sunny afternoon. Somewhat on the spur of the moment, Bill suggested that they should motor over to Arundel to take a boat on the river. For some reason that he could not entirely explain, even to himself, he did not mention the boating party the previous week. This time, he took a rowing-boat which was lighter and easier to manage. Even so, it was too hot for much purposeful rowing, but they paddled peacefully upriver, Betty holding the two steering ropes, and making gestures every so often to keep them out of the bushes.

They came to a shady pool where the water was calm and fish were jumping at unwary flies. Waterboatmen danced aimlessly over the surface of the water while a pair of blue dragonflies emerged from the reeds. 'What a lovely spot,' said Betty. 'It's so utterly timeless as to be almost unreal.'

'What a nice thing to say,' replied Bill. 'Yes, it's wonderful to relax and be at peace with life.'

Neither of them spoke for a while, then Betty's brow puckered slightly. 'Bill,' she said. 'When exactly do you leave for Ceylon?'

'In ten days' time, I'm afraid; I have to catch the Channel boat at Dover on the afternoon of the twelfth, spend the night in Paris and then take the express for Marseilles on the thirteenth. The boat sails that night.'

'Then what happens? Will you be coming back to England soon?'

'Oh dear, it's hard to look to the future. Provided there isn't a war, my contract with Liptons would be for four or five years, according to which estate I'm sent to.' After a pause, as their eyes met, he said, 'Betty, I wish I wasn't going so far away.'

'Me too, I've enjoyed our times together on your leave,' she said with a catch of emotion in her voice. Rather as an afterthought, she added: 'Will your mother be going back with you?'

'No, it has been wonderful to have had her with me, but now she is returning to live with her sister in Canada.'

'Then who will look after you?'

'I have a faithful manservant, we call them 'boys', by the name of Banda, whom I hope will be coming back to me.'

There was a slightly awkward silence which was broken by a pair of sculls as they glided downstream. Betty waved – acknowledged by a nod so as not to break the sculler's rhythm.

'Would you like to row for a while?' asked Bill.

'Yes, I'd love to,' said Betty.

He held onto the bank while they changed over. Above his head the dainty pink flower of the wild Alexander rose was swaying gently in the breeze. On impulse, he picked it, and, as Betty came to take his place, he put it in the clip to her hair. She smiled at him as she adjusted it.

Betty was a competent rower. There was a lake near to them at Bagshot, where some friends used to lend them a boat. Sitting in the stern of the boat, Bill watched her moving gracefully backwards and forwards as she plied the oars; like so much else about her, he admired the rhythm and co-ordination of her movements.

After a time, he took oars again. 'Betty,' he said when they had settled themselves again, 'before we left at lunchtime, I suggested to Dolly that we should all go out to dinner this evening. There is a nice restaurant in Chichester. I have booked a table for 7.45.'

'Oh, how lovely – then perhaps we should be getting back,' said Betty.

As they arrived back at the jetty, Bill did not see that Betty had carefully wrapped the little wild rose in her handkerchief and put it in her handbag.

The dinner in Chichester was a great success. Bill was glad that Betty was able to talk so easily and fluently with Dolly. Dolly, always a good listener, learnt much about the doings of the Francis family. At the end of the meal, Dolly announced: 'By the way, Bill, I nearly forgot. Jack Baker, your cousin at Crowborough, rang while you were out. He is sorry not to have been in touch during your leave, but he had two tickets for the Aldershot Tattoo on the evening of 9 June, which he cannot use. Would you like them?'

Bill looked at Betty. 'That sounds like us,' he said.

6

THE WHEEL OF FORTUNE TURNS

BETTY was enthusiastic about the idea of the Aldershot Tattoo. It was arranged that Bill should come to Park Hill for the nights of 9 and 10 June. After the Tattoo on Friday, they could spend Saturday together. Bill could return to Middleton on Sunday in time to get his packing done for him to leave on Monday.

Secretly, Dolly thought that this schedule was rather tight, but she did not say so. His packing only amounted to one trunk, which, in any case, would be done mostly by her!

Bill saw Betty off at Portsmouth station relieved that he would see her again. He would pick her up at the office on Friday after work. 'Have something warm for the evening,' he said as he kissed her on the cheek. 'It can be quite chilly after dark – even in June.' He watched her hand waving as the train drew away.

The next matter to deal with was the car. After giving the Austin a clean and emptying out all the bits and pieces that he did not want, he drove it up to London to the garage where he had bought it. He was expected. He need not worry – it would be on the quay waiting for him in Colombo. Then he caught a double decker for central London to have lunch at the Overseas Club. He looked around to see if there was anyone whom he knew, without result.

The train to Brighton seemed slow, and the bus to Middleton even slower. He was only just in time to pick up the car he had arranged to hire – a Singer 9 – before the office closed, but, thank goodness, he was mobile again. During the last few days of his leave Bill was restless and unable to settle. He played a few games of tennis at the Club, but did not stay on afterwards. Jan had rung up asking him to join an organised walk on the Goodwood estate, but he politely declined. Dolly noticed his disinclination and asked what was the matter.

'Oh, I don't know, Mum, I suppose my heart is half in Ceylon and half here.'

'You aren't thinking of not going back are you?'
'Oh, no. My life and work are out there and there I shall stay.'
There the matter rested.

Bill went to the estate agent to settle the rent of the house to the end of the month (Dolly was going back to the Langdales' at Compton until she left for Canada). He also called at the bank to make the necessary instructions about his salary. His travel tickets would be ready for him to collect on Monday morning.

On Friday he set off for his final weekend at Bagshot. The Singer was nice to drive and he now knew the way only too well. His restlessness was forgotten. Betty was waiting for him with a warm coat and a rug. Mrs Francis had thoughtfully provided a flask of hot coffee and some sandwiches. For some reason Jack Baker had left the tickets at the Queen's Hotel at Farnborough for them to collect before entering the grounds for the Tattoo. There was an enormous car park. They left the Singer to join the entering procession.

It was a long time since Bill had been in a crowd such as this; not, perhaps, since he went to the Tattoo at Earl's Court after passing his OTC exams. He found himself holding Betty's arm tightly, to prevent them being separated. In the growing dusk, the floodlights illuminated the huge arena. There was an air of expectancy and cheerfulness about the audience. A bombardier was manfully trying to knock in a support to the boundary fence which had collapsed; when his first attempt proved futile, loud cheers greeted his second effort. They found their seats. Bill hired two cushions as the chairs appeared rather hard!

The show opened with the massed bands of the Royal Marines and the Royal Air Force marching on from opposite ends of the arena – a remarkable feat of timing and of unison of step over such a distance. As the bugles echoed through, Bill fought off a nostalgic tear as he remembered his own participation when he marched through the City of London on St Matthew's Day. As the bands marched out, the stage was filled with the revving engines of the Royal Signals motorbikes with a daring display of balance, judgement of distance and timing. In one episode, the entire team of some twenty performers was perched, pyramid fashion, on five vehicles. Finally, one of them jumped, from a ramp, over a line of jeeps. Betty, who had not seen the display before, could hardly look.

For sheer strength and determination it would be hard to beat the Royal Navy Field Gun Competition, in which 4.7 inch guns with twelve pounder shells were dismantled and physically manhandled over an obstacle by seamen. The competition was won by the team first to reassemble, and fire, the gun on the opposite side of the 'ravine'. Cruelly, they then had to take it back again!

Bill's favourite episode was the musical ride by 'K' Battery, Royal Horse Artillery, in which teams of six horses and three drivers performed sequences of manoeuvres culminating in gallops across the centre, the horses missing the preceding gun carriage by inches. As the programme pointed out, there were no brakes in the event of misjudgment!

And so the evening passed to the final display in which detachments of the modern army dispersed elements of native tribesmen with much noise, smoke and zeal, but, bearing in mind the force of the attack, remarkably few casualties!

As they joined the queue at the exit, Bill felt a sense of pride at what he had seen – there was something invincible about the military presence of his country. As she held his hand in the crowd, Betty was a little subdued. Perhaps it was the sheer strength and raw masculinity of it all. Back at the car park, looking for the car, she soon recovered. For a time they hunted vainly for the unfamiliar hired car – then they remembered the coffee and sandwiches. They were in full view on the back seat. 'Here we are,' said Bill as he thankfully turned the key in the lock. Mrs Francis' thoughtfulness provided the end to a perfect evening. It was after midnight when they crept in through the back door.

The next day, Saturday, was firmly impressed forever upon Bill's mind. It was one of those beautiful English summer days which make one feel that it is good to be alive. Bill and Betty set out in the Singer intending to hire a boat on the river Thames at Wargrave. They made no direct attempt to get there but drove around the country lanes which were clad in the fresh green leaves of spring. Every now and then they were treated to the soft smell of bursting chestnut blossom or to the fragrance of viburnum or wych-hazel.

They came to a deep ford across the road with a wooden footbridge which Betty thought that it would be fun to cross. She got out to ascend the bridge and watch the exhaust pipe while Bill revved the engine before starting to cross. Betty yelled that it was being covered, but he did not hear – the car was through the ford.

At Wargrave they hired a rowing boat before taking it in turns to row lazily upstream waving at pleasure craft as they passed. They came to a hotel with a landing stage by the river. Hitching the boat, they climbed up to the lounge and ordered lunch. They were soon called to the dining-room, where the waiter referred to Betty as 'miss'. Then, after a quick look at Bill, he altered the 'miss' to 'madame', much to Betty's amusement. On the way back, they drew into the river bank to play some gramophone records, while Bill took some photos of Betty 'for him to take back with him'.

It had been such a wonderful afternoon – Bill would have liked to stay forever, but they had promised to be back for tea. Betty drove part of the way back after attaching her 'L' plates. Then there was a quick swim in the Pantiles swimming pool.

For their last evening Betty had booked a table at an old inn at Amersham called The Millstream. She had not been there herself, but had heard about it from friends. After a drive of about twenty-six miles, they arrived at the inn as the sun was setting over the town. Bill remembered how brightly the light shone on the red brick of the houses with their thatched reed roofs. The inn was an enchanting place. They entered through some heavy oak doors to climb from the first floor to a balcony, through which, behind glass panels, the millstream tumbled through the building. Searchlights, both above and below the water, caught the incandescence of the spray in a tumult of colour. Ferns nodded their graceful leaves at the rushing water, the sound of which echoed through the restaurant. They had a table overlooking the millstream.

They ordered the meal and danced dreamily to a radiogram on the polished oak floor below while it was being prepared.

'What a place!' said Bill. 'It's just too romantic for words!'

Betty flushed with pleasure. 'I thought that you would like it,' she said. 'I tried to find somewhere special for your last evening.'

'You certainly did!' replied Bill.

Returning to the table, Betty selected a bottle of Sauterne from the wine list, which the waiter then produced for approval. Betty was wearing the same silken dress that she had worn four months earlier at the Hohemunde, perhaps from nostalgia of that occasion. Bill noticed that the dark half-length cloak which had then covered her shoulders was now placed neatly on the back of the chair, leaving her shoulders bare.

The music continued. They danced between each course to spread out the enjoyment. Neither of them wanted to talk very much; they seemed to understand each other perfectly now, and Bill found his enjoyment beyond words. He sensed that Betty felt the same.

The coffee arrived. Bill had brought a box of chocolates and some Turkish cigarettes which she loved. He had a malt whisky before him while Betty had chosen a crème de menthe. They sat holding hands, listening to the millstream gurgling beneath them, watching the couples below and drinking in the surroundings.

Eventually, and reluctantly, they had to tear themselves away. There was no trouble about the car and they were on the way home, with Betty's guidance, through Beckonsfield and Windsor. In Windsor Great Park Bill drew into the side of the road. He put his arm around her shoulder and kissed her cheek. She turned her lips to his – it was their first real kiss.

The waxing moon was rising but the stars shone brightly in a cloudless sky. 'Let's see if we can find the Pole Star again,' said Bill. 'Yes, there it is above that tall oak tree.'

'And there's the Little Bear – *our* Little Bear – to the right of it,' said Betty.

For a moment their was silence between them, both, no doubt, thinking the same thought. Then:

'Bill, can you see the Little Bear from Ceylon?'

'Yes,' he replied. 'At least, the North Star is visible, though much lower in the heavens. If one can see the North Star, then the Little Bear must also be.'

'Then,' continued Betty, 'we could arrange both to look for it at the same time – it would be a sort of bond between us.'

'What a charming idea!' said Bill. He was laughing, but there was a sincerity in both their voices. He did not mention certain astronomical difficulties that crossed his mind, such as for example, the difference in the hours of darkness! For the moment, it was the thought that was all-important.

The next day, 11 June 1929, was the last day of his leave in England, but while he was with this family he tried to put that out of his mind. In the morning Betty and Jill took him for a swim at the Pantiles and they sat by the pool afterwards drinking fizzy lemonade. In the afternoon, Joan and her partner joined them for a four at tennis. Betty, in her white open-necked blouse and pleated skirt, looked so happy and charming. Bill once more found himself being unbearably aware of her presence. Her forehand was stronger than her backhand, so she took the right-hand court.

Came the evening. Betty announced that the goldfish needed some fresh waterweed, so she took Bill for a walk across some fields to a stream, where Bill, leaning precariously on the end of Betty's hand, retrieved some duckweed.

'Ugh, it's slimy!' he said.

On the way back, they passed a bank on the edge of one of the fields which was covered with wild white clover. They sat down for a moment in the sun. Close by, Betty spied a plant with tufts carrying four leaves. Picking two, she gave Bill one, keeping the other herself. 'That's to bring us both luck,' she said. Bill put his leaf carefully in one of the folds of his wallet.

Determined to make the most of the last few hours together, they went out in the car to find somewhere for supper – not very easy on a Sunday evening. Eventually they found a cottage advertising snacks and had scrambled eggs – a very generous helping, and coffee. Bill found that he was enjoying Betty's company more than he could describe; he wanted nothing more than just to be with her – it was an entirely new sensation. The thought of leaving her filled him with horror. Betty put her hand on his. 'Perhaps you ought to be going now,' she said. 'You have a long day tomorrow and you mother will be waiting for you.'

They drove around for a while and were approaching Bagshot. No, he thought to himself, we'll go round past that cottage again. Betty looked at him, but did not demur. Bill's head was in turmoil; something was now driving him which was beyond himself, his emotions were out of control. Suddenly he felt himself stamping on the brakes; they came to a halt on the side of the road. Betty looked back behind her anxiously, in case anything was following. Her face met Bill's arm on its way around her shoulders to grip her tightly.

'Darling, I love you. Will you marry me?'

She put her cheek to his. She had never heard him speak with such emotion. She did not reply for what seemed to him a lifetime.

'Oh, Bill,' she said eventually. 'I don't know. I love you too, but you must give me time.' A church clock nearby chimed nine times.

They drove back to Park Hill for him to collect his case. The whole family came out to wish him goodbye. This time he gave Betty a passionate kiss in front of all of them. Then he drove down the drive and was gone.

With Bill's final wave engraved upon her mind, Betty's self-composure melted. She ran upstairs to her room, shutting and locking the door behind her. Flinging herself upon the bed, she buried her face in the pillow to drown her sobs. Wisely, her mother left her alone.

For his part, Bill was in a dream, unable to relate himself to the world outside. Farnham, Hazelmere and Midhurst all passed in the gathering darkness without his being aware of them – the Singer was virtually driving itself. He was appalled at the enormity of what he had done. What could Betty possibly make of such a proposal? There was so much that he had wanted to say; to express his feelings for her; to say how much he loved her; to give her an chance to respond. To really hold her in an embrace. What an inadequate fool he had been! He now realized that he had been attracted by her since they first met at Lilian Killby's house. He must have been in love with her ever since Ivan had taken her out to the Sports Café at Seefeld and when he had felt so left out. Why had he left it so late to ask her to marry him? Now, at the very time of their lives when they should be building up their love and understanding, he was going nearly half the world away. Would he ever see her again . . .?

7

BACK TO THE WILDERNESS

It was nearly midnight by the time he returned to the cottage at Middleton. The light was still burning in the hall. Dolly rose from her chair as he entered the living-room and he suddenly realised that he had forgotten to tell her that he would be late. Poor Dolly; she had hoped to spend this last evening with her son and had prepared a special meal. Seeing the table still laid, he said with distress: 'Oh, Mum, I should have told you that I would be late!'

With typical self-restraint she replied, 'Never mind, did you have a good time?'

'Yes, of course I should have rung you. We had a wonderful time – they are all so kind. We got to the Tattoo and the next day Betty and I went to a lovely old Millstream restaurant at Amersham. . .' This reply was so incomplete. He and Dolly had always been so close, but how could he possibly say: ' . . . and I proposed to Betty and she asked for time to reply.' No, for the time being he must hold his peace.

Dolly was sensible enough not to press him.

Bill never knew how he managed to get off the next morning. There was the packing to complete, money and tickets to check and finally, the Singer to return to the garage. He gave Dolly his share of the rent of the cottage. She would return to Compton before sailing to Canada. She came with him on the bus to Brighton. They gave each other a loving embrace as the train left for Newhaven. There was quite a lump in Bill's throat.

During the Channel crossing he paced around the deck with his thoughts – it helped to pass the time. It was dinner-time when he mounted the steps to the Hotel Moderne in Paris. After a wash and brush-up in his room, he found the dining-room to be rather full. The waiter showed him to a table where a single man of about his own age was already being served with some rather sickly-looking tomato soup.

'Bill Baker,' he said, holding out his hand. 'Can I join you?'

'John Cunningham,' was the reply.

There was no lack of conversation. John was over in Paris for a trade fair where he was representing his firm from Manchester. Bill was to catch the night sleeper for Marseilles the next day *en route* for Ceylon.

Eventually, John said, 'What are you doing after dinner?'

'Well, nothing, really.'

'Then let's go out and see some of the lights of Paris!'

A slight pang crossed Bill's heart, but, 'Yes, let's,' he replied.

It was a warm, airless summer's evening as they stepped out onto the pavement with the light fading rapidly. The many-coloured street lights were already twinkling in the dusk, and the busy Parisian traffic rushing past them.

'Don't forget to look for oncoming vehicles on the right!' said John, with the air of a man who had been there before.

The brilliant red and white lights of the Moulin Rouge caught their eye, so they entered. An orchestra was playing soft music on a stage and the place smelt of scent and Turkish cigarettes. A waiter showed them to a table and Bill ordered a carafe of wine and canapés.

John was enthusiastic about his visit to Paris and proud to be representing his firm. He had some new electrical amplifying equipment for which he hoped to get some new contracts. After a while Bill found it all rather tedious, and the conversation one-sided. However, he noticed that the flow of John's repertoire was, at last, becoming distracted. He was looking over Bill's shoulder towards a girl sitting by herself at the bar. Turning his head, Bill could see that she was dressed in a long white silken skirt barely covering black high-heeled shoes. She wore a tightly fitting broderie anglaise bodice to which coloured sequins were uneasily attached. She smoked a cigarette from a long amber cigarette holder.

Without reference to Bill, John got up to address the girl. '*Mademoiselle*, he said politely, '*vous êtes seule?*' She nodded her head. '*Venez-vous à notre table?*' he said, holding out his hand. She accepted without, thought Bill, proper hesitation.

'*A boire?*' said John, rather casually.

'*Champagne, s'il vous plaît.*'

Bill had got up when she came, but then decided not to participate. He had heard tell of hostesses ordering champagne and receiving Perrier water.

'*Comment vous appelez-vous?*' demanded John.

'*Marie – et vous?*'

Moving closer to her and putting his arm around her: '*Où demeurez? Vous avez de logement?*'

Even the girl seemed surprised at the speed of his ardour. Bill got up to leave. This was the very last thing that he wished to become involved with.

'*Bonsoir, Mademoiselle,*' he said politely, nodding his head. 'Bye John, good luck,' he added as he walked away.

Bill spent a restless night beneath his duvet. The room seemed airless and there was no fan for ventilation. Besides, the bed was too soft . . .

After a leisurely lunch on the Boulevard he collected his luggage and caught a taxi to the station. He had a better night in the sleeper lulled by the gentle movement of the coach. A quick breakfast on the train before they pulled into Marseilles. A taxi to where the P & O *Strathallan* lay, silent and mighty, in her berth. He soon found his cabin and dumped his baggage on the bunk. Before unpacking, he made his way forward to the staff quarters below the bridge. Without hesitation, he opened a door marked 'Private – Staff Only' and walked along the corridor until he found another door marked 'Wireless Operator'. Knocking, he walked in.

'Good morning, Cyphers,' said Bill with an air of a man carrying out his everyday business. 'Are you still connected with the shore?'

'Well, yes sir, but . . .'

'Then could you send an important message to England for me?'

'There are Regulations . . .'

'This is a matter of life and death,' said Bill, purposefully.

'Well, in that case, sir . . .'

Bill handed him a cablegram that he had written on the train.

> To Miss B. Francis,
>   Park Hill, Bagshot, Surrey, England.
> From Baker, S. S. *Strathallan*, Marseilles.
>
> Darling I love you.
>
> Bill.

The wireless operator read the message and gave an exaggerated wink. Bill left the room hurriedly, leaving an English pound note on the table. Returning to his cabin, he unpacked and sat down to write her a most passionately loving letter to be posted before the boat sailed at midnight. He ended by asking her to write to him c/o the boat at Bombay, realizing what a dreadfully long time ahead that would be.

The boat was full. As usual, a group of unattached young people established themselves for a round of games on the deck, swimming, dancing and generally enjoying themselves. Bill saw no reason why he should not enter into all this, but recounted his doings to Betty in his next letter posted in Port Said.

In June the Red Sea was blisteringly hot. There was very little daytime activity. Even in the evening, most people were content to sip iced lemonade in such breeze as there was on deck. At Aden there was the usual scramble by the boat-side traders – Bill bought a string of amber beads for Betty which he would send with other presents that he had in mind, for when he reached Colombo.

They were all glad to reach the Indian Ocean – even though it became rough. Bill, normally, a good sailor, had to spend a couple of days on and off his bunk.

As they neared Bombay Bill felt a growing excitement at the thought of receiving a letter from Betty. Impatiently he waited at the Purser's office for the mail to be sorted. To his joy, he was handed an air mail letter in her firm and distinctive hand. Clutching it he took it down to his cabin to read, dreading that he would be refused. He knew that she would soften it as much as she could, but would conclude that the whole situation was utterly impractical. His hand shook a little as he tore open the envelope.

It was a most loving letter. She remembered several little happenings during their time together which had been important. She would never forget their evening at the Millstream – she had wondered whether he was going to propose to her that night! In the event, he had left it so late that there wasn't time to discuss some of the more practical issues of life affecting a planter's wife. She enquired about the effect upon the health of children, their schooling and further education. She was keen that their offspring should follow some sort of professional career in England.

A feeling of elation spread through Bill's mind. He had not been refused. The questions were what every girl would want to know, and he loved her for thinking of them. It struck him how blind he himself had been in this respect when choosing a career in Ceylon. He immediately scribbled off a note to her in loving terms, saying that he would answer as soon as he got back to Pooprassie. In the meantime she was never out of his thoughts. He handed this to the Purser, making sure that it would catch the post in Bombay.

Bill felt a strange feeling as he turned onto the estate road leading up to the Big Bungalow at Pooprassie. So much had happened to him since leaving six months ago; no doubt things would have changed here too. One of the most immediate issues was to find out whether Banda would come back to him.

Brazier opened the front door to him with a mixture of courtesy and formality. 'So you are back.'

'Yes, sir, and it's nice to see Pooprassie again.' He followed the Peria Dore into the drawing room, where Mrs Brazier, remaining seated, held out her hand.

'You had a good leave? You must tell me about it sometime. There is a young lad named Cox, who has just joined the company, in your old Bungalow – you will be in the Senior Assistant's Bungalow by the factory, and will be in charge of that Division. Unfortunately, this is being redecorated at the moment; you are welcome to stay here for a day or so until it is finished.'

'Thank you very much,' said Bill, addressing Mrs Brazier as well. 'I should like to.'

'Peria Tini will be at 8.00 – Rajasami will show you to your room,' said Brazier, pressing a bell by his chair.

Bill was shocked at the appearance of Brazier. He had always been thin, now he seemed skinny. He walked like a man uncertain on his feet. Wisely, he had made no comment.

After unpacking his washing bag and pyjamas, he set off down to the factory to find his tin trunk with the clothes that he would need for the morrow. The factory was a hive of activity – there was a flush of leaf following the break of the monsoon and yesterday's leaf had withered late. Tea-making would not finish until late tonight.

Mr Patel, the Teamaker, greeted Bill with a toothy smile. 'Salaam master, I heard that you are back. You will be coming to the Lower Bungalow?'

'Yes,' said Bill. 'So it seems that we shall be seeing quite a lot of each other.'

They talked for a while in English before returning to his room at the Big Bungalow. Today, he reckoned he was still on leave. Tomorrow he would be at 6 o'clock muster and would meet his new Head Kangany.

With an hour or so left before dinner, Bill went back to his room at the Big Bungalow and started his letter to Betty. It was not easy to portray his love and longing to be with her, and, at the same time, to answer her questions about married life in Ceylon. He did his best to describe the climate, the planter's life on and off the estates, the bringing-up of children and their later education and the opportunities for wives to travel to and from England. On reading it all through several times, and trying to look at it through Betty's eyes, he realised, perhaps for the first time, how much he was asking of her. He ended by saying:

> There are a few matters that are rather more in my favour; Ceylon is one of the most, if not the most, beautiful places upon earth. If you were to come out here you would see the world while you are young enough to enjoy it. You would have a home and garden in lovely surroundings, servants and freedom for club life if you want it. People are very keen on tennis, and though you might not think it by my play, the standard is really quite high! You would have entertaining to do as a Manager's wife is expected to host brass hats who are sent up from Colombo. Wife's passages for Home leave are paid for by the Company every five years Up-country or four years Low-country, but, of course, there is nothing to prevent you going Home in between leaves for special needs. (Though I should miss you terribly!)

By the time he had finished (with another session after dinner) he had covered eight pages of manuscript.

He ended:

> All I can now say, my dearest Betty, is that if you ever did decide to come out to marry me, you would have, not only my everlasting love, but also my unsurpassed admiration as well.'

Bill closed the envelope and held it a while in the palm of his hand. It was, perhaps, the most important letter that he would ever write. Opening the door, he went in search of the houseboy whom he found on his hunkers outside the kitchen. He handed him the letter and two rupees, asking him to tell the tappal coolie to post it by air mail. He should be called at 5.30 a.m.

The muster ground was on the forecourt in front of the factory. Bill arrived on the stroke of 6 a.m. to find Periasami, the Head Kangany, waiting for him with all the other kanganies lined up to be introduced. Bill was touched by this little piece of courteous formality. As he went along the line he put his hands together as each name was given to him, and which he did his best to remember. He was relieved that Periasami appeared to be a man who was clear and precise in his orders and spoke to Bill in well-pronounced and grammatical terms which he could easily understand.

To Bill's joy, there standing at the side of the muster ground was his horse, a white Arab stallion with a long flowing mane, 'Regent' by name. Kistnan, the horse-keeper, at Regent's side, bowed to his master, while Bill fondled his mount lovingly.

Turning to Periasami, and in Kistnan's hearing, Bill explained that on this, his first day, as he and Periasami had so much to discuss he would like to walk with him round the estate during the morning. He would meet him at the factory at 8 a.m. He would need Regent during the afternoon.

Bill was at the breakfast table when Mrs Brazier came in. Bill got up and held the seat for her. She was glad of the opportunity to explain how poorly her husband had been. The previous Senior Assistant had left to take an appointment with another company and Michael Cox on the Middle Division was inexperienced. She was thankful that Bill was back. Brazier was having treatment in Colombo.

Bill and Periasami had much to talk about as they went from one gang of coolies to the next. At this stage Bill was more concerned to learn the caste and relationship of the various kanganies with their charges, and indeed, the extent to which the caste system operated within the Division. One kangany in particular had been over to India recently and had brought back four families to work on the estate. This had given him considerable 'status'. On the whole, Bill was pleased with what he saw, though a certain amount of 'tail-twisting' seemed to be required on the extant '*kanics*' (tasks) meted out to the gangs.

Bill was eager to have his first ride on Regent and took him up to the plucking fields during the afternoon. Now that he was a Senior Assistant, he was not expected to keep the checkrolls himself – there was an English-speaking conductor to do that. As soon as the leaf had been weighed in the evening, Bill went into the factory to see how Mr Patel was getting on. The day before had been so wet that the poor man (and the night shift) had been up most of the night trying to get the leaf withered before the tats were needed for this day's leaf. The fans had been working all night and in the end he had had to apply heat as well – much to his reluctance. However, now all the leaf was down and the rollers were pounding away twisting the leaf.

It was nearly dinner-time before Bill arrived back in the Big Bungalow;

he had to scramble in and out of the bath. Brazier offered him a sherry. Bill had had a good day and he was glad to be able to talk about it.

It was not until after dinner that Bill was able to give thought to his own affairs. Priority now was to contact the family who had taken Banda when he went on leave. The arrangement had been on a permanent basis, but Bill dreaded the thought of starting with a new boy. He had kept the address and rang the Superintendent of an estate over in Dimbula.

For once, luck was on Bill's side. The family had been planning to go on furlough to England, but owing to the situation at Home they had arranged to go to South Africa. They were sailing next week and Banda was free to come back to him. Bill immediately wrote out a cheque for Banda's train fare and a month's salary in advance.

Then there was Bunty. Bill rang his friend at Haputele and a cheery voice answered. 'Yes, he is fine – come and collect him when you wish.' Bill would come over at the weekend.

Next day Brazier himself took Bill down to the Lower Bungalow to have a look over it. Whoever it was who was supposed to be looking after John Digby must have had some pretty weird ideas. The whole place needed to be turned out and scrubbed from top to bottom and redecorated. Bill had his own linen packed away in the factory, but new bedding was required and some of the furniture needed mending. For once Brazier became quite apologetic. The estate carpenter would be sent down, the place would be cleaned and new bedding ordered. By the time that Banda arrived the work was almost completed. Bill moved in two days later, and the following weekend collected Bunty from Haputele.

Bill heaved a sigh of relief when it was all done; in the end the Braziers had been quite companionable, but it had been a strain living with one's boss! He was delighted to see Banda again who immediately enquired after Lady. Bill decided to increase his salary to Rs. 35 and a half bushel per month. Now that he was a Senior Assistant's boy, he had a status to keep up!

As for Bunty, he was overjoyed to see his master. He had had to play second fiddle to a rather large and boisterous Red Setter for too long. Bill did not dare to tell him that the Setter was soon coming to stay with him!

Two days later, the South-West Monsoon broke. For the last day or so it had been hot, dry and airless – almost eerie in its atmosphere of expectancy. People had been known to take off all their clothes and run round the garden in the sheer joy of feeling the cooling water running down their naked bodies, but Bill resisted the temptation! Brazier had rung up at midday to say that work should be broken off against a full day's credit, and Bill was glad to regain the shelter of the Bungalow. Bunty looked like a drowned rat!

In the evening, after he had seen Regent fed and made sure that he had been properly rubbed down, Bill sat down to write to Betty.

My Darling Betty,

Outside it is raining – raining with all the uncontrolled fury of the onslaught of the South-West Monsoon. I can hear the roar of the water as it pours down a ravine, not eighty yards from the Bungalow, from whence it cascades over the Irish drain on the cart road, before falling into the river in the valley below. Brazier has ordered that all work should stop and, after luxuriating in a bath, I am now sitting in front of a huge log fire in the grate and listening (when I am not thinking of what to say to you) to the wireless. His Lordship, Boots of Bandara Eliya, commonly known as Bunty, is curled up in front of the fire, dreaming of the jungle cock that he put up yesterday, and altogether it is very cosy! My greatest wish would be that you were here to share the enjoyment . . .

Bill went on the describe the 'bachelorish' nature of his drawing-room and to anticipate what Betty would want to do with it. He guessed that most of his school and rugger groups would be for the chop, and that the plain blue hessian curtains might not meet with her approval. Anyway, they would have the greatest fun replanning it all.

This led him on to contemplate upon his hopes and aspirations for their marriage, concluding that the all-important factor was that they should truly love one another and support each other in the ups and downs of life that were bound to occur. He had no doubt in his own mind that he was, not madly, but soberly and lastingly crazy about her. Secondly, he longed for her companionship to give him a purpose in life: he yearned passionately for her company beside him and her support in everything that he did.

It was a very loving letter, again several pages long. He could not wait until the tappal coolie brought another letter from her.

But there were other things to be done. Bill had received a letter from the Adjutant of the Ceylon Planter's Rifle Corps asking for a report on his attachment to the Rifle Brigade in Winchester. The Adjutant wanted to know whether this had been of benefit to him, and whether further attachments should be arranged. Bill had also been made Hon. Secretary of the local branch of the Ceylon Rifle Association. There was a range not far from Kandy – meetings had to be arranged.

The rain had stopped. Bunty had woken up and was demanding attention. Bill called in at the factory in order to listen to Mr Patel's worries over wet leaf. Then he paused for a while at the bridge over the river to watch the weight of water surging below, carrying in its path rocks, debris, and even whole tea bushes. The water, grossly discoloured by the erosion of soil that went with it, frothed as it whirled. How long could this level of erosion of the best soil continue, Bill wondered.

Bill received two letters from Betty that week: one written in reply to his posted in Port Said and the other following his written in Bombay. The cable that he had given to the wireless operator in Marseilles had caused great excitement. The telegraph boy had brought it while they were all having breakfast. Betty had opened it, not without anxiety, and had blushed as she read it. Led by little Ruth, they had all wanted to know what was in it, but Betty folded the cable and slowly returned it to the envelope. 'Bill has reached Marseilles and boarded the *Strathallan*,' she said a little flatly. After breakfast, she went up to her room gripping the envelope tightly. Lifting her blouse she placed it in her bosom.

Betty concluded the second letter by thanking Bill for the detail in which he had answered her questions about life in Ceylon. 'Bill,' she had written, 'before we decide to get married, we *must* be certain of our real and lasting love for each other; only time will tell. In the meantime, the sensible thing seems to be to carry on with our normal lives. If our love is true it will prevail over all else.' Poor Bill! She was right, of course. He was thrilled by even the reference to their being married.

His life continued too. One morning he had ridden Regent up to the pruning field. Kuppen, the Kangany, was already on the path and came over to Bill to say that Appavu, one of the coolies, wanted to have a word with him, and beckoned the coolie over. Appavu must have come up from behind Regent's flanks and touched them as he passed. Regent's reaction was immediate – prancing forwards then backwards, he sheered off the path to get his hindquarters enmeshed in one of the tea bushes at which point he and Bill parted company. Bill picked himself up while Kuppen retrieved his topee from the middle of another bush to the general mirth of the assembly.

BACK TO THE WILDERNESS 131

It was not often that they had the chance to see the Dore biting the dust! After examining Regent's hoof, Kistnan, the horse-keeper, made to rebuke Appavu, but Bill intervened. *'Kariam ille, Appavu,'* he said, *'alun kuthere oru pindi kuda vara.'* (Never mind, Appavu, but never come up to a horse from behind.)

Poor Appavu. In all the commotion, he had quite forgotten what it was he wanted to say to the Dore in the first place!

That weekend he went down to the Club – the first time since returning from leave – and received a great welcome. There were many enquiries after Old England and how everyone was shaping up to the prospect of war. Then there were the inevitable personal questions. Was he . . .? 'No!' said Bill, emphatically, though somewhat wistfully. He played two sets with Sarah Jordan, who had also been teasing him about his leave.

'You've got a strong forehand today, Bill,' she remarked as he hit a ball with particular vehemence.

Later, instead of the cushion game, Bill organised a .22 shooting competition on the Club range behind the factory. He had arranged for the Adjutant of the Ceylon Planters' Rifle Corps to come to the Club the following month to take part in a shoot, and wanted some practice, both for himself and the rest of the team.

The month of August was passing quickly. Everyone was getting preoccupied with events at Home. He was now receiving a weekly letter from Betty which arrived on Thursday. The tappal coolie was asked to deliver this personally to him wherever he might be in the fields. It was interesting that the bush telegraph was such that the coolie had no difficulty in finding out where the Sinna Dore had gone. Sitting on the nearest rock, with Bunty at his heels, he would tear open the letter to read its contents avidly.

Next week he received a cable from his mother for his twenty-fifth birthday. She had arrived in Canada. Betty also sent him her best birthday wishes and love.

On 3 September he heard on the overseas broadcast Neville Chamberlain's message to the nation. They were at war with Germany.

It seemed that plans had been laid well in advance. The CPRC was still under mobilization from the 1938 'scare'. The next morning he received a telephone call. He was to report at Trincomalee at 16.00 hours on 6 September with arms and equipment. He was on duty with an advance party to pitch camp at China Bay. Rendezvous China Bay Station. Remainder of the Kandy Company would follow on 8 September. So he was off. Brazier

Water From the Lake

was not best pleased. He was talking about getting Bill exemption on the grounds of his own ill-health.

Bill was among the first to arrive at China Bay along with the Company Quartermaster Sergeant. Their task was a repeat of the previous occasion. Tents had to be erected and other stores drawn, latrines dug and provisions brought in. It was hard work. He was thankful when the officer ordered them to stand down. They stripped naked and ran into the sea.

After dinner the officer drew them all together. The wireless station was part of the East and Far East communication system that was vital to the war effort. It was a soft target which the Germans might make a pre-emptive effort to destroy. This might be by a landing party further down the coast or by a landing at the port under cover of a bombardment from the sea. In either case the CPRC was to provide ground protection for the station. The remainder of the company, when they arrived on the 8th, would be given further orders.

When the rest of the company arrived there was much activity renovating some of the old positions on the hills above the bay, digging new ones with better fields of fire, setting up command posts and laying field telephones. It was all a light-hearted affair – few thought that their Lee Enfield .303 rifles and Lewis guns would have much effect against modern armour, but at least they were 'in the war'! Bill was concerned that they seemed to be digging in for a static engagement, when the more likely intention of an enemy would be to get in quickly, destroy the objective and get out. He discussed it with the Platoon Commander.

'You are trying to run before you can walk, Corporal Baker,' he said. 'When these positions have been prepared you will be required to reconnoitre the coastline, firstly around Koddiyar Bay and later further south and North. We shall have to be familiar with possible landing sites and routes that the enemy might take to the Naval Station.'

Bill gave a grunt of approval. 'Thank you, sir,' he said. The officer was on an estate in Dimbula and was well known to him. He played centre three-quarter for that team. Before his accident, Bill had brought the officer down within yards of the Uva line. Now, things were different!

And so it was. They spent the next fortnight wading through jungle, swamp, salt-marsh and river. They mapped the approaches to Trinco in great detail. They got covered in leeches, were footsore and many were covered in prickly heat as the result of their exertions, but the job was done. They had a daily dose of mepacrine against malaria and salt for the prickly heat.

They had been in Trinco for nearly three weeks. At morning parade the CO announced that the Kandy Company would be returning to the estates in two days' time, after having been relieved by another company. He went on to say that it was not the intention that the CPRC should man the base at China Bay indefinitely. A regular battalion would be taking over that duty in due course. The corps would continue to divide its time between the estates and military training, both of which were of national importance. The news was greeted with applause; most of the company had wives and families at home.

Bill arrived back at Pooprassie to find two letters from Betty awaiting him. He had only been able to scribble codified notes to her from Trinco. She was greatly concerned as to whether, as the result of the war, Bill would be coming Home or not. While he was in Trinco, the War Office in England had made a proclamation that all holders of the OTC Certificate 'A' should notify the Office giving the number of their certificate, and their whereabouts. On returning to the estate, and after searching his old papers, Bill sent the information saying that he wished to enlist in England. However, at the same time Ceylon Defence Headquarters issued a statement that the national interest required serving personnel to remain with their units in the island.

Poor Bill. What could he say to Betty? He told her of both these statements. His firm, Liptons, had made it clear that if and when their staff were called up for service outside the island, they would make up any shortfall in their pay and hold their seniority. This would not apply if they left by any other means. It seemed doubtful, bearing in mind his attestation in the CPRC, that he would be able to get away. After agonizing over the question, he decided to tell her that, at the moment, there seemed little chance of his getting home. It was a dreadful letter for him to write – he hoped and prayed that it would not make any difference to their love. He received no reply from the War Office about his Certificate 'A'.

Bunty was overjoyed to see his master back. Banda had been taking him for walks, but this wasn't the same. Kistnan had been exercising Regent by leading him around the estate, but this didn't satisfy him either; he was raring to be ridden. For his part, Banda had used the time to make the Bungalow spotless. The *dhobi* had washed the curtains and Banda the carpets. There was a new floor in the kitchen. Bill was pleased with the result, and told Banda so.

Three days after he arrived back from Trinco Brazier rang him. 'You are to go to Dambatenne,' he said.

'Yes, sir?' said Bill in disbelief.

'All the Assistants have been called up to Trinco, and Prior had to go on jury service. The estate is empty. He wants you there tomorrow.

As Bill left Pooprassie he carried with him the expression on Bunty's face when he realised that he was being deserted – again! The journey through the hills and mountains took him four hours to cover the eighty-eight miles, such were the bends, hairpins and gradients. He stopped for a coffee in Newara Eliya, the hill town at 6,200 feet. It was a pleasure to see English flowers growing in gardens and yellow gorse on the banks.

Prior greeted him. 'I'm glad to see you, Baker,' he said. (It was nice to have a name given him, even though it was a surname! Something that was lacking at Pooprassie.) 'I was afraid that I would not see you before I had to go.'

He handed Bill a day sheet of matters in hand and then continued. 'As a matter of urgency I would like you to go, straight away, to the estate boundary by Monrakandy where a heath fire has been raging since this morning. Some of our tea is threatened. My horse is in the stable and Ramasami is expecting you.' Finally, gathering his papers together he said, 'Mrs Prior and Monica are away, but Suppan, my Head Boy will look after you. I hope to be back next week.'

Ramasami was, indeed, expecting Bill who mounted Ranger and set off for the scene of the burning heathland.

After dinner, Bill found a corner in the Big Bungalow by a reading lamp and sat down to write.

Dambatenne,
Haputele,
Ceylon.

30 September 1939.

My darling Betty,

Have you, I wonder, ever pictured Charles Lamb as he wandered through deserted quadrangles and courts at Magdalene during the Oxford Vacation? The walks were so much his own; the halls empty, doors open to invite one to slip in unperceived. To play the gentleman and enact the student. In moods of humility, to imagine himself a Seiser or Servitor; in those of ambition, a Master of Arts.

If such a mood has ever held you, Betty, you may find it easier to understand my feelings at this time.

Here I am, the only European on this huge place, the greatest in the Company with some 2,200 coolies and 1,400 acres together with a perfect palace of a Bungalow, charged with the responsibility and direction. A position which I cannot hope to occupy for the next 15 to 20 years. The charge is mine of the next five days! Real as it is, there is a wonderful sense of exhilaration, and of course, Challenge . . .

Forgetting the war for a moment, Bill's letter continued in a new vein of purpose and prospect. There was a hope for life in the future and an enjoyment of the present which, perhaps, had been lacking in his earlier letters.

The next day he was greeted by salaams and smiles from the coolies, most of whose names he knew, and was able to enquire after their families. Many asked after 'the Lady'. Although she could not speak their language many regarded Dolly as something of an oracle! His greatest thrill came when he passed a ten-acre clearing which he had reclaimed five years previously and which he had contoured and pegged himself before planting. It was now in full bearing. Joining a gang of pluckers upon it he immediately recognised the Kangany.

'Salaam Sinnamuttu,' said Bill with a smile.

'Salaam, Dore. Welcome to Baker Dore's Patch.' So he had a place in history!

Sinnamuttu was a mine of information. Kadiaveil, the Head Kangany, was becoming very elderly and now seldom came out to the fields. Bill made a mental note that he should find time to go into the lines to see him. Muttusami, the old stonemason, had sadly died – a natural death, he added quickly. He had not blown himself up! There had been great celebrations when Kadiraveil's grandson Arian had married Kaliamma, daughter of another high-caste kangany. The drums had been beating all night and everyone had drunk too much arrack.

The days passed quickly at Dambatenne. As he was the only European he could not go off the estate as it was a company rule that at least one should always be present. However, one afternoon he arranged for some old friends to come for tennis. The houseboy brought them tea in the pavilion. It was then that he heard some news which caused him great sadness.

While he was on Mousakellie, Bill had become friendly with an Assistant on an adjoining estate. Robin Davy was about two years older than Bill, but they shared common interests. Although Robin had never been in the Uva rugger team, he had been a reserve on a number of occasions, so Bill had given him a lift down to Badulla. They had also been on jungle shooting trips together and once had stayed at the rest-house at Hambantota for a boating trip. Bill had regretted that, since his move to Pooprassie, they had lost touch.

It seems that Robin had gone on leave at the end of the previous year and became engaged to a very attractive young girl. She came out during the summer when they were married at Haputele church. It was quite a big wedding and local social event. They seemed an ideal couple and were both obviously happy. They went off to Newara Eliya for their honeymoon.

Two weeks after they returned, a young Sinhalese woman and her father appeared at the Bungalow. She was carrying a newly born half-caste baby which, she said, had been fathered by Robin. There was a scene during which Robin, losing his head, tried to 'buy her off'. At that moment his wife appeared. Quickly sensing the situation, she ran screaming from the Bungalow to seek refuge with the Superintendent. She had returned to England without seeing Robin again, and he had heard nothing from her since.

No one knew how the matter would be resolved, or even whether the child was Robin's, though most people seemed to think that it probably was. He had been a social hermit ever since. The episode left a lasting impression on Bill's mind. How a man's life – and for that matter, a woman's too – can be ruined in a matter of seconds. There was some banter and eye-twitching in the pavilion which accompanied these revelations, in which Bill did not join. Robin had been and, as far as he was aware, still was, a friend of his.

Paddy Prior duly returned from jury service and Bill was glad to report on all that had taken place. They went up together to see the result of the fire. On Bill's direction, much of the debris had already been removed but the damaged tea seemed beyond redemption. The police sergeant had been as good as his word and supplied the name of the owner of the land. Prior said that he would see whether a case for damages could be brought.

His short 'act' on Dambatenne over, Bill expected to return to Pooprassie but Prior had other ideas. There was no sign of the Dambatenne Assistants returning from Trinco, and, as Brazier had Michael Cox, Prior rang him to ask if Bill could remain with him for the time being. 'No,' said Brazier, 'I want Baker back.' There had never been too much love lost between the two Superintendents. Not to be outdone, Prior rang the firm in Colombo. Baker could stay! Bill gained a certain satisfaction from being fought over!

Bill spent most of a day, the next week, sitting outside the Court at Bandarawela waiting for Prior's claim for negligence against the owner who caused the fire. He used the time to write to Betty. The case – which was lost – over, Bill left for the journey to Pooprassie.

As he was driving down the road to Haputele, he turned, on a sudden impulse, up the cart road to Robin Davy's bungalow. He action was entirely unpremeditated; he wasn't sure what reception he would receive, or what he would say. He had been slightly aggravated by the conversation in the tennis pavilion. It was obvious that his friend was in deep trouble; perhaps he would like some company and, possibly, sympathy.

He knocked at the front door and, after a pause, a boy answered.

'This is Baker Dore,' Bill said. 'Is master in?'

'I will see, master, please wait.'

There was a further pause, then Robin appeared. Bill was shocked by what he saw. Robin had aged by at least ten years; his eyes, previously bright and dark with an easy smile, were leaden and lifeless. His fingers were deeply stained with nicotine. He looked at Bill unsurely and inquiringly.

Bill held out his hand. 'Hello, Robin. I've been to Dambatenne for a few days. I thought I would look in on my way back to Pooprassie. Can I come in?'

'Of course,' said Robin, leading the way.

They sat down in a rather sparsely furnished room. Robin poured two whiskies. They looked at each other, unsure of what to say. Bill, feeling it incumbent upon him:

'I miss the trips we used to have when I was over here. Do you remember when we went to Hambantota and went fishing for mullet? I think that our 'catch' was only four fish for the day's effort!'

A glimmer of recognition crossed Robin's face. It was hard work, but Bill continued with the recollections. Eventually: 'I don't leave here much nowadays. You've heard that my wife has returned to England?'

'Yes, Robin, I have heard that. You must miss her terribly.'

'You've heard why?' Robin continued.

'No,' said Bill firmly. 'The tropics can be a shock to a young girl when she first comes – the change in the environment and the heat. Now I suppose that you are separated by the war. I just came to say how sorry I am.' Robin seemed satisfied to leave the matter there. Bill rose to leave soon afterwards. 'You must pick up some of your old interests,' he said, shaking hands again.

He went out the way he had come. In the meantime, ostentatiously, as if to establish a territorial right, a pram had been placed on the verandah.

Bill had plenty of time to contemplate Robin Davy's plight during his four-hour journey. One thing he was absolutely clear about was that, come hell or high water, he would never let himself get into the same position.

Waiting for him at Pooprassie were no less than three letters from Betty. These made him feel nearer to her than at any time since he left England. The letters were sweet, loving and affectionate. They all wanted to know when he was going to come Home. Regretfully, Bill felt that he should quote from a broadcast made by MacDonald and which appeared in local newspapers entitled 'The part to be played by the Colonies'. It ran thus: 'To all those thousands who have applied from Overseas to serve with our Army, Navy and Air Force at Home. I must point out that their first duty is to ensure the defence of their own land and the strategic bases contained therein upon which depends the strength of the Empire. I can assure you that everything is being done here to ensure that the manpower of the whole British Nation is employed in the most useful way and that everyone will find themselves in that capacity.'

He continued: 'Well, Betty, these are high-sounding words, but I am afraid that it means that I shall not be given a passage Home. Patience seems to be the only thing that we can rely upon at the moment.' He did not tell her that someone at the Club had told him that, in response to his own efforts to get Home, had been told that, in the event of releases from Ceylon being made, Singapore was a more likely venue. Then Bill responded meticulously to Betty's letters. It was after midnight before he put down his pen.

Bill came back to Pooprassie to face trouble with the Tamil labour force. The estate had never had the happy atmosphere that prevailed on Dambatenne. There had been an undercurrent of tension between two different castes that Bill had noticed when he first arrived on the estate. Bill had discussed this with Periasami, the Head Kangany, whom he judged was not involved. The rivalry stemmed from jealousy between the principal kanganies of two lower castes. The difficulty was that both these men had brought a number of families over from India to work on the estate, all of whom remained loyal to their chief. Bill had mentioned this to Brazier and had sent him a confidential note on the subject, just in case.

The present trouble arose over the treatment of a woman of one caste by a middle-aged man of the other, whom, it was alleged, he had sexually assaulted. The man, whose wife and family had left him to go to another plantation, denied the case against him. He was quite well educated and Brazier had given him a minor clerical job in his office. Bill suspected that therein might lie the cause of the trouble. After a long investigation, and with the help of Periasami, Bill found out that the real cause of complaint was that some of the paysheets had been altered to the detriment of coolies in the other caste. Obviously the clerk had to go, not because of the assault, but for having tampered with the books. Bill had the invidious task of telling Brazier about it all. He wasn't too pleased about losing his clerk!

However, there was a happy ending to it all. Bill knew the Superintendent on the estate where the wife had gone. After a little tail-twisting the wife agreed to have the man back, and he left the estate to join her. Bill later noticed much improvement in relations between the two castes.

Over a weekend, Bill had been invited to join a young party to watch the Pera Herya – the Buddhist ceremony of the parading to the Tooth in Kandy. The party gathered at the Queen's Hotel where a balcony had been booked overlooking the main street. There was already much excitement down below them; the streets were lined with expectant faces of many Eastern nationals: Sinhalese, Tamils, Burmese, Malaysians and even Sikhs and other northern Indian races. All were dressed in their traditional vestments, the women in their brightly coloured saris, their dark hair brushed flat and shining with coconut oil. Elaborately fashioned jewellery in gold hung from their noses and ears, contrasting with the silver buckles and brooches that adorned their hair. By contrast, the men were simply dressed in white jackets and loincloths, their teeth stained red with betel which was spat periodically onto the pavement. To lighten their attire some wore coloured turbans with silver or gold edging. Most were in bare feet, but those who considered themselves 'Babus' had leather sandals. The lights flickered above the chattering lines of humanity; the smell of the East wafted the air – a mixture of frangipani, woodsmoke, betel, shuruttu (cheroot) and coconut oil. Then the tom-toms in their weird, uneven rhythm, slowly at first, but gradually working up to a crescendo of captivating, almost stupefying noise. The crowd was silent now, their faces lit by a parade of torch bearers, their staves fuelled by burning and smoking oil, waved plumes of fire on high. Then followed the Kandyian dancers in historic garb, gyrating in time to the beat, the bells in their headress and their ankles picking up the drama of the exotic rhythm. Bare to the waist and waving fire sticks, the men seemed to exemplify the ancient ritual of the procession.

By way of contrast, there followed the child dancers, some no more than tiny tots, dressed in red and white, led by a young Sinhalese beauty, but keeping perfect time to the tom-toms as they danced. They received much applause.

Another expectant pause – and then the elephants covered in regal drapings of satin laced with gold and silver thread with tassels hanging beneath their bellies. These huge animals, emblems of strength and power, moved slowly and gracefully, their long tusks tipped with gold and their mahouts resplendent in white and gold. Some seventy elephant passed sedately, each interspersed with more tom-toms and pipes, when at last in culmination, the largest elephant carrying a brilliantly lit basket containing the object of the whole procession – the casket with Buddha's Tooth. In its wake followed the High Priest himself, a splendid figure with great solemnity.

After the procession had passed there was silence for a while. The rhythm of the tom-toms, the spontaneity of the dancing, the majesty of the elephants and finally, the symbolism of the Tooth had left an unforgettable impression, to which words could make but little addition.

Afterwards they went the bar at the Kandy Club, where tongues were soon loosened. It was late by the time that Bill reached Pooprassie.

It was now nearing the end of October 1939, and the second month of the war. Whatever was happening on the East European Front, the involvement of the Western Allies seemed a little unreal and, perhaps, even more so, to those in Ceylon. Nearly all the members of the CPRC had returned from their initiation at Trinco and were now rigorously engaged in local training, mostly over weekends.

It was now four months since that fateful evening in June when he had proposed to Betty. Every detail was implanted upon his mind. 'Oh, Bill,' she had said, 'I don't know. I love you too, but you must give me time.' How many times had these words run through his mind? He still had not had her reply. However loving her letters were, she still ended 'Yours ever, Betty,' while his . . . well, he just spoke his mind.

It was 2 November. Bill was up in the pruning field discussing with Kuppan, the kangany, the task for the day. Two of the pruners were sick which meant that the others had to work extra hard. Bill was concerned that the work might be rushed. Regent and the horse-keeper were waiting on the path. Regent, impatiently, was flicking some horse flies from his flank.

Out of the corner of his eye, Bill had seen Pichi, the tappal coolie, leave the factory and start to make his way up the hill towards them. So there was a letter for him from Betty. He resisted an impulse to leave the field and go, running down to meet Pichi, holding out his hand for the letter. However much he wanted to, such would hardly be seemly for the Sinna Dore. Instead, the tappal coolie slowly climbed up to them. After groping in his bag, he handed Bill the letter.

With propriety still in mind, Bill pocketed the letter, and, dismissing Kuppan with a nod, mounted Regent. He also turned to the horse-keeper. 'I will see you back at the Bungalow,' he said, urging Regent into a canter. Bill made for a favourite spot of his, where a stream surged and tumbled over rocks and there was plenty of shade under an old grevillea tree for Regent and himself. Dismounting, he tied up the horse and sat down to read Betty's letter. It was an especially loving one in which she reflected that which he had written from Dambatenne, full of hope and courage for the future and, determination to made their love survive. She quoted something that Bill had not previously noticed – the motto to the crest on the letter heading of the P & O *Strathallan*. *Quis Nos Separabit* (Who shall separate us?)

As the pages turned to the final paragraph she ended 'Yours lovingly,' but it was a small neatly written postscript that sent a thrill like an electric shock running down his spine. It ran –

'Darling, if at *first* you don't succeed . . .'

A feeling of ecstasy ran through his whole being. Jumping upon Regent he cantered down to the Bungalow. Throwing the reins to the horse-keeper, he drove in a dream down to the Post Office in Galaha.

His hand was shaking a little as he wrote out a cablegram:

Betty Francis, Park Hill, Bagshot England.

Darling I repeat my question of 11 June with all my heart.

Bill.

He handed the scrap of paper to the Sinhalese postmaster. 'Thank you, sir,' he said. 'That will be five rupees and four annas.' Never had Bill handed over the money so willingly.

It was forty-eight hours before he had a reply. The longest in his life. He had given a rupee to the postmaster and asked him to ring as soon as there was a cable for him. The telephone had rung several times the previous evening. Twice it had been Brazier and once the Teamaker. At last, the following evening, 4 November, the postmaster rang.

'Mister Baker, sir? There is a cable for you. Shall I read it?'

'Yes, please,' said Bill, hardly able to contain himself.

'It reads like this: "Baker. Pooprassie, Galaha, Ceylon – "'

'Yes!' said Bill. 'Please go on!'

'"The answer is yes darling may God bless us both. Francis." That is all, sir, I will send the copy up in the morning.'

'Thank you,' said Bill as he rang off.

Bill felt like going out into the garden and shouting – shouting and throwing up his arms to Heaven. Of jumping onto Regent and galloping, galloping anywhere until the horse dropped. Of ringing up his friends and saying, 'I'm engaged!' But the mood passed and real life returned. Instead, he went into the bedroom and knelt by the bed, as he had done as a child. He prayed that God should, against all the odds, bless them both.

As the realisation of his engagement began to sink in the original sense of joy and elation became tempered by a new sense of his responsibilities. Gone would be the heady days of self-expression and self-interest; his life was now in trust to Betty and hers to him. Clearly she must be a party to any major decisions, and there was the very practical issue that he should start saving for their marriage. He had a wonderful and glowing feeling that he was now no longer alone: even though she was so far away, Betty was a part of his life. Perhaps it was that, even to himself, he had not admitted how much he needed the love, affection and sense of belonging that she could give.

With another CPRC training session threatening, there was much to be done. He tried several times to put on paper the emotions that he had experienced from her cable. Finally, he concluded that no words that he could muster could really convey his feelings and love for her. Perhaps it was sufficient to say so in his letter. He wrote about her ring – what were her favourite precious stones? What type of setting would she like? – what a shame that they could not choose together.

Then he wrote to Mr Francis asking for his daughter's hand in marriage. This he did not find easy; he really had not very much to offer his future bride, except himself. Even that carried a high degree of uncertainty in a far-off land. Not for the first time did he realise how much he was asking of Betty. Yet Love transcends all else.

It was two weeks before he heard from Betty following their engagement. It would be hard to imagine a more loving and affectionate letter. They had

had a family party to celebrate. Mr Francis had made a little speech about seeing the family growing up. He had given his blessing to them both. Further time went by before he heard from her about the ring. She left the choice entirely to him, but perhaps something with diamonds or sapphires would be nice.

Bill immediately rang Lewis Siedle, the famous jewellers in Colombo, to make an appointment for the following Saturday. Like most young men, his knowledge of precious stones was nil. Mr Siedle was most patient. They would make the platinum shank themselves and recommended a large square-cut sapphire from the mine at Ratnapura. Mr Siedle waxed eloquent in his description of the variety of stones laid out on a green satin cloth before them. The stone should be flawless, and uniform in colour and brightness. Two supporting baguette diamonds would reflect light through the sapphire to increase its brightness. These would have been mined in South Africa. After earnest deliberation, Bill chose the stones and the setting.

'Thank you, sir,' said Mr Siedle. 'By the way, what is your fiancée's ring size, please?'

Bill's heart sank. He had no idea! Bill would have to cable and let him know as soon as possible. Bill wrote out a note to go with the ring: 'To Betty to seal our everlasting Love, from Bill.' He enclosed the note in an envelope and gave it to Mr Siedle.

The ring arrived safely in England. Her praise was beyond words. In return, she sent Bill a photo of herself wearing it.

As he had expected, Bill found himself back at Trinco. A garrison defence for the radar station and harbour had now arrived, but the local defence was still assigned to the CPRC who were considered to be used to the climate, and familiar with the topography. It was intended that the Corps should be the first defence against an attempted land-based assault.

This official statement caused a certain amount of ribaldry in the mess, particularly about the climate, which was hot, damp and sticky – unlike anything on the estates. But that was the job to be done. This time they were in the jungle to the north of China Bay between Irrakkakandi and Pidavkaddu, astride lower reaches of the river Kal Aru. The only crossing of the Aru was by ferry at Salappal, which was an obvious strategic point to be defended against a landing north of there.

As an exercise, they were divided into 'Blue' (Bill's side) and 'Red', in defence. Having established their position, 'Blue' were to plan a route to China Bay. During the following day the platoon commander asked for volunteers. There was only one obvious track from the beach up through the jungle. Reconnaissance had established that the 'Red' troops were in the locality. The officer wanted a volunteer to swim round to the rear of the position to see whether there was another way out.

Bill volunteered. Stripped to the waist and with gymshoes he swam across the bay to another promontory. It was now dusk and he could move about more freely. There was no one. He was now further inland and made to swim back to report. He was swimming evenly and without hurrying. Suddenly there was a 'plop' on his right and a movement on his left. An icy shudder ran down his spine. Crocodiles! [3] He looked behind him – there was movement in the water there also.

His only course was to swim ahead making as much disturbance of the water as possible. Thrashing the water with all his might he swam faster that he had ever swum before. Reaching the shore, he scrambled out – the evil-looking creatures were staring at him.

He reported to the platoon commander. They took the main route to the jungle outlet where the 'Reds' were waiting for them. Most of the 'Blues' thought that this was the intention of the exercise anyway!

By the end of this period of training, they were tired, footsore, bitten by mosquitoes and leeches, sunburnt and suffering from prickly heat. Some of the scratches sustained in the jungle had, despite rigorous medical inspection, gone septic. Of the many lessons to be learnt, self-care was high on the list.

As they left, they all held a vision before them – that of a long, long lie in a steaming hot bath!

Bill returned from Trinco to find several letters from Betty awaiting him and also a registered packet which contained a gold signet ring inscribed with his initials. He received the ring with sheer joy; it became a talisman of her love – something real to hold. An everlasting bond denoting the trust that they had given to each other. Sometimes, in moments when he needed her most, he would put his right hand over to touch the ring and imagine her presence.

The letters were preoccupied with the question of how they could get married. It seemed to be less and less likely that Bill would be released to

---

[3] The Ceylon (Sri Lankan) Crocodile is a slightly smaller version of his African cousin. It is six to eight feet long and quite agile. It can be a vicious customer, especially when hunting in a pack.

return to England, and he still had had no response to his letter to the War Office, offering himself for a commission following his Certificate 'A'. On the other hand, as the war situation became increasingly grave, the thought of asking her to come out to him became more and more agonizing.

Christmas came with no apparent solution to the problem. Bill and Vincent were invited to lunch at the Big Bungalow, at which the Braziers were excellent hosts. Brazier held forth with his reminiscences, and Mrs Brazier played the piano. As soon as the two young men could decently leave, they went up to the forest pool to swim and lie in the sun.

Bill was becoming quite a connoisseur of precious stones. For Christmas he had sent Betty an Indian rupee silver brooch set with beautiful blue-green transparent aquamarine stones which he had found at a jewellers in Kandy. These stones had also been mined in Ceylon and cut and polished by native craftsmen. Then, for her birthday in January, Mr Siedle had made for her a pair of white gold earrings set with sapphires and moonstones. All arrived Home safely; Betty was unstinting in her love and praise of them. This gave Bill a new interest. He had always been fascinated by the adornments worn by the Indian women. Ceylon was a honeypot for both precious and semi-precious stones and the native jeweller's craftsmanship.

In March there were rumours that Bill was to return to Dambatenne. One of the Assistants, who had been on leave after Bill, had tried to join up at Home. His application had been refused (no doubt partly at the intervention of the firm) and he had been told to rejoin the CPRC and his job. He was coming back to Pooprassie.

At the same time he received a letter from his Aunt Jane. He opened the letter with some surprise: they had always exchanged Christmas cards, but he had not had much contact with her of late. Jane had seen the announcement of his engagement in *The Times* and sent her love to them both. She thought that Bill would like to know that her niece, Betty Laurie, whose address she enclosed, had just become engaged to a tea planter named Laurence Carey in Ceylon. She thought that it would be nice for the two girls to meet.

This opened up a new dimension. Bill knew Carey slightly: he owned an estate on the other side of the island, so Bill decided to contact him. 'Yes, by all means put the girls in touch.' How confusing that both of them were called Betty!

Bill heard from 'his' Betty that she had been invited to stay. She was looking forward to the visit.

Then Brazier rang. Harry Carter, the Assistant on leave, had arrived back in Colombo. Bill was to hand over to him next week and then go back to Dambatenne, where he would be in the Senior Assistant's position. Whoff! A feeling of elation spread through him at the thought of returning to his old estate, but he would have liked more time to settle up.

Harry duly arrived. He had been on one of the last boats to sail through the Mediterranean. It was probable that future sailings would be via the Cape. He painted a rather grim picture of life in England with the blackout and rationing. He appeared not to want to talk too much about it and Bill respected his reticence. He had left his wife and child at Home.

Bill's next letter to Betty was from Dambatenne and was a memorable one, 'I'm so happy to be back here,' he wrote. 'It really seems that I have come home. This is a fine new bungalow with a fine garden. Were it not for my insatiable yearning for you, I should be really contented. If you were here, darling, life would be as near perfect as upon anywhere else on earth.'

As he was writing, Banda, who was still gasping over the speed of the move from Pooprassie, came in with a letter from Betty. Bill tore it open. It seemed that she and Bet Laurie had got their heads together and were thinking about coming out together! If so, they would be sailing in about six weeks' time. It is hard to describe the joy that filled his whole being. His love for her welled up within him, as if his emotion would burst. He read the letter, reread it again and again – it was really true.

In the next few weeks there were frantic preparations. Betty with her trousseau, her permit to travel, and her passage. Bill sent a draft to his bank at Home with her fare. He thought for a while that he should suggest a double wedding with the Careys, but then decided against the idea. Theirs was likely to be a social event that he could not afford. Instead, he asked the Neishes (she had just returned from England) whether Betty could stay there until they were married. He thought that they would have a simple church wedding in Colombo. He discussed this at length with the padre in Haputele.

Bill's joy and anticipation knew no bounds. But in his more sober moments he knew how much he was asking of Betty. Not only did he feel a sense of guilt that it was she who was facing the danger of the journey and not he himself, but he was asking her to leave her family and be married in a strange land away from them, to a man whom she had met on his short holiday. He had no illusions about the extent of her commitment. He loved her passionately for it.

Prior had been helpful and understanding throughout. Bill was careful to see that his work did not suffer. It was a long day from the time that Banda called him at 5.30 until he had finished in the evening and written part of his weekly letter to Betty. His horse at his new Division was 'Skittles', a magnificent English hunter who needed to be mastered and understood. Boots had not altogether approved of the move; he had settled into his ways at Pooprassie and had a tacit unison with Regent. Now he had to be more careful of Skittles' heels! Still, he had the advantage that he was allowed into the Bungalow, where he was on his own devices. On the few occasions when master sat down in an easy chair, he usually ended up on master's lap.

It was now the beginning of May. Bill had done everything he could think of for Betty's reception and their wedding. He still did not know whether they would be going round the Cape, a journey of some eight weeks, or through the Med in three. It might well be that she would ring up and say 'I'm here!' Bill would stop in his tracks at such a thought, half-closing his eyes. It was unbelievable.

He had been to the muster ground followed by a talk with the Head Kangany and a visit to the factory. As usual he had come back for breakfast. On the table was a plain buff-coloured envelope. A cable from England! he tore it open. It read:

> Mummy and Daddy in car crash, Mummy serious, Daddy injured. Fear cannot leave them. Utterly disappointed. Betty.

Bill put the cable into his pocket. He went out onto the verandah and gazed at the valley below. Fate had played a cruel trick. but at least Betty was all right. For an awful moment, while opening the envelope, he had feared otherwise.

It was a fortnight before Bill learnt the details. Mr and Mrs Francis had been driving home after dark following a visit. It was during the blackout. Mr Francis had driven straight into the back of a stationary lorry which he did not see. The car had gone under the tail with such force that the windscreen had collapsed upon Mrs Francis who had been scalped. They were both severely concussed. Mr Francis had been beside himself with remorse and regret. His wife was still in hospital.

Poor Betty. She had lost her passage. For the moment she did not know what to do. Bill had already written offering his greatest sympathy and love. There was not much that he could do either.

Bet Laurie duly arrived out in Ceylon. Bill was invited to the wedding but, feeling that he could not face it, sent his apologies and a wedding present.

It was obvious that Betty was going through a very bad time. The after-effects of the accident, the anti-climax of her passage, the worsening war news and, indeed, Bill's uncertain future, were having a marked influence upon her. Putting aside his own dreadful disappointment, he did his best to support her courage and faith in the future. However, when a month later a cable arrived suggesting that they should postpone further attempts by her to get out, he agreed, but disguised a heavy heart.

Bill threw himself into his work, which became progressively more demanding. An undercurrent of labour dissatisfaction was developing in the island — some said communist-inspired. One tea planter, who had had a mob stoning his Bungalow, opened fire upon them. Together with some other planters, Bill had been called out as a detachment of the CPRC 'In Aid of the Civil Power' to a disturbance at Pussellawa. Luckily their presence was enough to enable the police to take charge.

As the result, Bill had to keep his ear to the ground about what was going on at the estate. At the first sign of a coolie being *wrange* (disobedient) Bill would insist that he and the Kangany would get to the bottom of the trouble, however long it took. The policy paid off, and the happy name of Dambatenne was maintained.

One day, when making his way to one of the more remote parts of the estate, Bill decided to take a short cut, on foot, through the jungle. He told the horse-keeper to take Skittles by a longer way around on the bridle path. The jungle path, little used, was overgrown with bamboo, creeper and thorn through which Bill had to force his way. Boots looked at his master disapprovingly. As they progressed, they came upon a clearing where a stream gushed from a granite rock face above, falling to a pool before them. It was hot and steamy. Streaks of light filtered through the leaves of the grevillea trees above, casting moving shadow upon the running waters. Green and blue dragonflies darted hither and thither, alighting on the yellow and orange Lantana bushes. A tree frog landed with a plop into the steam. There was an air of tropical majesty.

Boots, who was now ahead, came to a sudden halt, his ears cocked and his whole body alert.

There upon a rock protruding from the middle of the pool was a pair of cobra courting. Bill froze and watched, fascinated. The pair, their hoods erect and their wet shining bodies glistening in the dappled light, were swaying gently side to side as if in time to some unheard rhythm. The male, larger than his companion, would now and them break the rhythm by lunging forward as if the bite her face. She would respond. Then the swaying would resume. Gradually, he entwined his long frame with hers until they were lying together in the shallow water. Then they were still for a while, until the male, having achieved courtship, swam to the edge of the pool before disappearing into the undergrowth. Returning to the rock, the female curled herself up contentedly to sleep.

Bill was deeply moved by the scene. If these creatures, which God had designed, could fulfil their love so gracefully and tenderly, what cruel fate had come between him and the one he loved? And how long would that love be denied? He stood for some time, engrossed with his thoughts, until Boots reminded him that it was time to move on.

In common with other planters, Bill was now spending nearly half his time with the CPRC either at Trinco, or some other remote part of the island, or on training exercises at Diyatalawa which was the Up-Country Depot for the armed forces. He had been nominated by the CO at the top of a list of NCOs to be trained in a cadre of instructors in the use of small arms. When the list got to Headquarters in Colombo, his name was taken off the list on the grounds that he was likely to go overseas shortly. When he heard this, Bill immediately wrote to enquire what this signified, reminding that he had applied for release to join up in England. This resulted in the Ceylon Defence Force referring the matter to his firm. Doudney replied that on no account could he be released. It all seemed stalemate again, and in the meantime, Bill had lost his cadre!

At the end of one of these exercises, Bill found himself, with his car, near Laurie Carey's estate near Badulla. Weighing in his mind the heartbreak at the cancellation of Betty's passage against his desire for any news that Bet Carey could give him about Betty, he decided to ring up and ask if he could call. They invited him to dinner. Bet painted a vivid picture of Betty's desolation, both her parent's accident and her disappointment at having to cancel the passage. Bill hadn't realised how ill Mrs Francis had been (and still was). Betty's letters had described vividly what had happened, but hearing it again from Bet brought the whole matter alive. Bet said that they had become the best of friends over the arrangements for the passage. Bill was a very lucky man to have such a charming fiancée. For his part, Bill had a greater realisation of what his Betty had been through. He would reflect this in his letters. He was glad he went to the Careys'.

Following the cancellation of Betty's registration for a passage, most of Bill's efforts were again directed towards getting Home. The firm still flatly refused to release him and the Ceylon Defence Force said that his first duty was to stay in the island. If, however, the situation altered, his release would certainly not be to England. It seemed a complete stalemate. It was now over a year since they had seen each other.

Bill was pleased to hear that his cadre had been reinstated. He was to report to the Echelon Barracks in Colombo for at least two months. Another Assistant, who had been on leave in South Africa and who had married while he was there, would return to Bill's Division and Bungalow. Bill would return to the remote Division of Mousakellie; Boots would have to go to friends.

Still, it was nice to get away; the training which was arduous and realistic helped to provide another interest. He had always been good at musketry – now he would prove it! There were sea landing exercises, laying mines both on land and underwater, the use of grenades and practice with the new Bren gun which had only just arrived in the island. Unlike some of the others, Bill enjoyed the bayonet practices where he could let off some of his pent-up frustrations!

At the same time there was a new determination about Betty's letters. She now seemed convinced that Bill would not get out. She was thinking of re-registering for a passage! With the previous disappointment, he scarcely dared hope. Then a cable: 'Sailing shortly terribly excited. Francis.' He responded: 'Wonderful news. So am I.' Bill immediately went round to Cook's the travel agents to find out what they could tell him. The boat was the *Staffordshire* and would have about a hundred people for Ceylon on board. It would be sailing in about ten days' time and would be out about eight weeks after that. Bill was not a little surprised at the amount of information given to him! His joy, love and pride in his fiancée knew no bounds.

Back in the barracks, he put everything he had got into the cadre. In the final pass out he came top with eighty-three per cent and was allocated a 'distinguished' grade. He had congratulations from the CO and, believe it or not, from Liptons. He was asked to stay on in Colombo as Instructor to the following cadre.

It seemed, at last, that events were turning their way. Bill could imagine Betty's frantic preparation for leaving, of the poignancy of her goodbyes, her journey up to Liverpool and the *Staffordshire* waiting for her, Her life would be so much in the hands of others; not his. It was a dreadful decision for her to have had to take. He just loved her for it.

But for now, the new cadre was arriving. He was a temporary Sergeant once more!

The first week of the new course had passed. Then Bill received a cable:

Sailing instructions mislaid in post. Passage missed. Bitterly disappointed. Francis.

Bill felt numb all over. As soon as he could leave his duties he walked out of the barracks and along the seashore. There were tears in his eyes. But he must stop being sorry for himself. He turned around and made for the Post Office. His response read:

How cruel on us both. So sorry for you. I love you. Bill.

At the end of the course Bill returned to Mousakellie where Banda was pleased to see him, and Boots, when he was fetched, was utterly delighted. Bill tried to set his mind on other things, but he was beginning to feel that his life was being manipulated by some outer force that was beyond his control. Poor Betty – when her letter arrived it was clear that she was overwhelmed with grief and a sense of anti-climax. After weeks of correspondence with Cook's Travel and having been told that she was booked on the *Staffordshire*, the final and vital document which the agents said that they had posted never arrived. Apparently all the other passengers had received theirs, which gave the date for joining the ship.

Bill did his best to comfort her – he would have to live with his own disappointment. She did not say anything about another passage, and Bill did not dare to raise the subject.

Soon after Bill returned to Mousakellie there was an outbreak of typhoid in the coolie lines. The patient had recently arrived back from a visit to southern India and must have brought the infection with her.

In common with other Europeans, regular inoculation was a 'must', but Lipton's did not normally insist on this for the labour force. However, Prior decided that everyone should be injected. A young Ceylonese doctor arrived and asked Bill to assist. His job was to hold the arm of the patient, apply iodine and wait for the doctor to make the 'jab'! There were five hundred coolies: it took the whole day. Most had never had an injection before. Their reactions varied from stolid obedience to outright panic. Strangely, many of the children were so overcome by being touched by a white man that they didn't notice the prick!

Bill had played rugger with the doctor, who came up to the Bungalow for lunch. He was in no hurry. They spent a long time on the verandah chatting and drinking fresh lime and soda. Inevitably, they got onto the question of Home Rule for the Ceylonese. While posturing politeness itself and acclaiming British Rule, he predicted that India would become independent first, soon to be followed by Ceylon within, say, the next ten years.

'Do you indeed?' said Bill. 'Then what will happen to the tea estates after that?'

'Well, I suppose that they will be nationalised,' was the reply. Bill did not take this view seriously, but it was the first time that he had heard it given with such conviction.

The next week the company was in Diyatalawa. Bill was one of five instructors whose job it was to get each man through his 'efficiency' tests. The task was formidable. On the whole, Bill was thankful to have his mind fully occupied. The members of his squad were not above getting their instructor under the table in the bar during the evenings. His powers of resolution were not only confined to musketry training! At the end of the session, the CO called the instructors in to thank them – everyone had passed the tests.

The year of 1940 was drawing to a close. Bill had sent off presents to Betty, and to his mother, and many Christmas cards as well.

He renewed his efforts to obtain his release to join up in England and even obtained an interview with the Brigadier of the Ceylon Defence Force, who promised to look into his case. Agonizingly he waited for the formal reply. It was as before. His duty was to remain with his present Unit. If his release were to be granted in the future, it would be either to the Middle East or to India.

Bill wrote to Betty. He was beginning to feel that they must resign themselves to the effect that their marriage during the war was improbable. But Betty had other ideas. She had made a further application for a passage out. Unspeakable joy and love for her again filled Bill's heart, but, with the worsening war situation, could he, indeed *should* he, allow her to take the risk? Once she arrived what security could he offer her? Here were two young people madly in love with tragedy and uncertainty all around them, clutching at straws of happiness. Betty's determination prevailed. She reapplied for a passage.

The uncertainly of waiting was awful. Surrounded by wedding presents to bring out, and putting final items in her trousseau, waiting for the daily mail was agonizing. At last it came. She was to be at Liverpool to board the *Strathallan* on 20 March 1941. Bill could only picture the departure – the elation, the tears the anticipation and the realisation that she was on the footing of a new world.

This time she was on board. She was sharing a cabin with two other girls, one going to India to be married, the other a nurse going to Trinco. Sailing round the north coast of Ireland at night, they could see the lights on shore. In the bar, a young lad joined them. 'Robert,' he said holding out his hand. He was a medic, going to Burma. On the way to the Cape in South Africa, the *Strathallan* was taking a wide berth out into the North Sea.

After breakfast the next morning the three girls went out onto the boat deck. It was cold, but they found a sheltered corner for their deck chairs. Robert came to join them, sinking somewhat luxuriantly into a fine cashmere rug. It was a bright, clear morning, perhaps a little too fresh for a spring day. The water was calm. As far as the eye could see there were no other boats in sight. Everyone was chatting happily, Betty, particularly, thrilled to be at last on her way, was proudly telling the others of her expectations in Ceylon.

A speck appeared on the skyline, watched carefully from the bridge.

Full speed; Alarm; Boat Stations; Man the Gun; ordered the Captain. It was a German Fokker dive-bomber.

Everything happened so quickly that Betty could not remember where she was. Someone grabbed her as the ship suddenly lurched violently to starboard. They were close to the anti-aircraft gun which members of the crew had already manned. At boat stations they were ordered to lie on the deck. The plane made a pass over the ship, accompanied by the crack! crack! crack! of the Bofors. Slowly, and in its own time, the plane turned for a second run, this time dropping a clutch of bombs two of which hit the ship with a sickening thud followed by an explosion. One had pierced the engine room, bringing the ship to a halt and starting a fire. For some merciful reason the aircraft did not press home its attack but made off. With the fire gaining hold and fearing a further attack, the Captain ordered the passengers to abandon ship.

Betty was in one of the first boats to leave. As the crew were letting down the gantry, Robert flung her his rug; a simple act of kindness which may have saved her life. She wished that he had got onto her boat, but apart from male members of the crew all the occupants were women. They saw some other boats get away, but when a fog came down they lost touch.

They drifted in the open sea for four hours. Fours hours of nightmare; one of the passengers became hysterical and had to be injected to make her sleep. The others sang songs and, under the eye of the officer in charge of the boat, issued out some of the hard rations. Betty was getting very cold, although everyone huddled up together. Then they sighted a ship. The officer sent up a flare – they had been seen! Someone said a prayer of thankfulness which everyone joined in. It was a Norwegian cargo ship whose crew cheerfully hauled them aboard.

They were set ashore at Stornoway in the Outer Hebrides. News of their plight had come before them and there was quite a crowd on the jetty. In the Town Hall a doctor checked them all before they were led away, individually, to residents' houses. Betty was given a hot meal, a hot-water bottle and a warm bed where she slept eight hours.

When she awoke the family were at breakfast. She found that they had even lent her a nightdress and a dressing-gown – she had no recollection of getting into bed. Never had she been so thankful to be alive and for the kindness of these people. She rang home with what was a highly emotional call.

Mr Francis sent Bill a cable. 'Betty shipwrecked. Lost everything. Safe. Home Thursday. Francis.'

Bill's reply was awaiting her arrival home. 'Thank God you safe. Love you for ever, darling.'

Poor Betty. She arrived home dishevelled and disheartened. She was also not a little shocked. She was still clutching Robert's rug. Other things had been borrowed from the Health Authority. Later, it emerged that the *Strathallan* had not sunk. Most of the crew had stayed aboard and put the fire out. She had limped into port under her own steam. Betty's luggage was safe. Sadly, one of the boats had capsized on being lowered from the ship with many of the passengers drowned. A list would be published later.

Bill, alone up at Mousakellie when the cable arrived, felt an appalling sense of guilt at having allowed her to sail. The more he thought about it the worse it became. He sat down and wrote her the most admiring, loving and adoring letter had had ever written. He was now forever in her debt for what she had done for him. Now, somehow and from somewhere he would get Home to marry her.

This third and final attempt to join him in Ceylon must put the seal on their marriage out there. Their love must now be kept alive until he got Home, possibly until the end of the war. He had no delusions about the enormity of the challenge.

It was some time before he heard from Betty with the details of what had happened. There had been great sympathy for her in the village and everyone had been so kind. The local paper had run a front page article about her experiences, but she could have done without such notoriety. Her letter was full of sadness and concern for Bill. She had been in touch with Cook's to find out Robert's address in order to return the rug. She wrote to the address given, only to hear from his parents that he had lost his life when the boat capsized. They suggested that she should keep the rug.

Bill was sent to act as Superintendent on Panilkandi while the regular man was away, only to find that he was returning early, making a wasted effort. Then he had to go down to Dambatenne again while Prior was away. Bill was beginning to experience an entirely new feeling of disenchantment with his job. Lying on his bed one night he remembered the words of the little Ceylonese doctor. 'I suppose that, after the coming of Home Rule in ten years or so, the estates will be nationalised.' What if he were proved to be right . . .? It was strange how, quite suddenly, things seemed to have become '*deja vu*'.

Bill was thankful to get away again on military exercises. There was to be another episode with HMS *Norfolk* the battle cruiser on the Near East Station. This time it was to be a counter-attack upon the positions established by an 'invading force'. They were to receive air support by airlift and supply. They would have to penetrate deep into the jungle where maps were imprecise. There was room for quite a lot of local decision and enterprise. The *Norfolk* landed them on a remote and windswept beach and left them to it!

Back on Mousakellie there was plenty to do. Bill spent the first two days getting to grips with what had been going on during his absence and discussing cropping programmes with Prior. As the result of the last ten days' exercises and his efforts on his return, Bill was tired, almost exhausted. As usual, he had seen to the feeding of the horse and had had his bath. Banda had brought him dinner, and, after clearing had said: 'Good night, master.' This evening Bill had replied:

'Good night Banda, thank you for all that you have done while I have been away. The Bungalow looks spotless.'

Banda beamed. 'Thank you, master,' he said closing the door.

Bill collapsed into the chaise longue, a whisky beside him, writing paper in his hand. 'My darling Betty,' was all that he could write before the pen fell from his hand. He was asleep.

He was awakened by a knocking at the door. It was Banda looking uneasy and distressed. Bill frowned. This was an unusual and unwarranted intrusion.

'I'm very sorry, master. There is a man here who wants to see you. I have tried to send him away, but he won't go. He insists on seeing you.'

'Right, I'll see him,' said Bill, reluctantly climbing out of the chair before following Banda out to the rear passage.

There, standing in the passage, was a Tamil man of late middle age, a cloth loosely tied around his head, bearded, with his teeth stained with betel. A patterned woven shawl hung from one shoulder above a stained white shirt and *veti*. His legs were bare save for cheap leather sandals. By his side was a young girl, no more than sixteen or seventeen, her head covered by a headcloth the side of which she had pulled across her face and covered her mouth, as is the Indian practice when in the presence of a high caste. She wore no bodice, her sari hanging from one shoulder partly exposing one of her breasts. Round her waist was the traditional blanket or *kumbly*. As Bill looked at her she moved backwards, cowering behind the man.

'Salaam Dore,' said the man, putting his two palms together and bending forward in a bow. Bill did not answer. Continuing in Tamil: 'Master want nice clean young girl for the night, only fifteen rupees?' As he spoke he pushed the girl forward roughly, pulling the headcloth from her face and the sari from her shoulder, baring her chest. 'I come and fetch her tomorrow,' he concluded, wagging his head from side to side.

Bill felt himself going white with rage. His body began to quiver with fury. Turning to Banda, he was about to rebuke him for having allowed this to happen, but the poor man was so crestfallen that Bill just managed to control himself. Instead he turned to the pimp.

'You come, after dark, to the Dore's house. You invade my privacy uninvited. You come with the sole purpose to ply your vile trade. You have no thought for the dignity of the person whom you choose to address and you resisted my Appu when he told you to leave.'

'Now take this poor girl away and treat her with respect that she deserves. Get out of my house, and if I ever see you on this estate again I shall call the police.'

Re-entering the drawing-room, he slammed the door. Grabbing the whisky glass, he gulped down its contents in an attempt to assuage his anger, pacing the room as he did so. Still quivering, he went out onto the lawn in front of the bungalow. With all the love, hope and disappointment that had been so much a part of his existence, what demonic power could have introduced this incident tonight? Surely the devil itself, he thought.

Then, strangely, a curtain seemed to draw itself over his mind. He looked up towards the heavens. The moon was shining but the stars also shone brightly. There was the Plough, the Great Bear, and just visible above the horizon – the Little Bear. He knelt to pray. The tears were streaming down his face.

With the spread of the war into the Middle East the air mail service was suspended. All letters had to go via the Cape, and took about six weeks. Even that service became erratic. An alternative route was tried via the Far East and America called the 'Chunking' service but this was also later withdrawn. Bill had written several letters to Betty for her to pick up at the Cape, but had no means of knowing whether she had received them. Those letters had been so full of love and expectation that to receive them in England would only cause pathos and grief. He had sought to rectify this in later correspondence but did not know whether this had helped.

News came that a boat carrying a major shipment of tea to the United Kingdom had been sunk. An order came that all estates were to increase production immediately. In spite of the seriousness of the order, it caused a good deal of mirth in the planting world. Some wag wrote on his paper 'Wrongly addressed – refer to the Almighty', sending a copy to the local newspaper. Bill discussed the matter with Prior. There was absolutely no waste of leaf anywhere. They could, perhaps, delay pruning a while, but this would have the effect of reducing yield in the longer term. They decided to ignore the order.

Then things began to happen quite suddenly. One of the company's Superintendents, who had been on leave in South Africa, returned unexpectedly. Doudney was prepared to release Bill! Almost at the same time, the Defence Force decided to send a contingent to the Officer's Training School at Bangalore, India. Bill was given three weeks' notice to leave. Liptons would keep his job open with seniority and would also make up his pay to his civilian level.

He had a wonderful send-off from the Haldumulla Club where he and Johnston, another member, were toasted in champagne. 'Johnnie', in typical Cossack style, draining his glass threw it over his shoulder to smash on the floor. Not to be outdone, Bill poured the dregs of his over his pal's hair.

All the estate kanganies turned out to see him off and Banda, to whom Bill had given three months' salary, had tears in his eyes. The final sadness was his parting with Boots. The poor little chap seemed to know that it was the end. He had a good home to go to with one of Bill's friends.

At Colombo station, from where the contingent left, the entire firm, including Mr and Mrs Doudney and Wilmshurst, the Deputy, turned out to see Bill and an office worker, Mackie, off to India. It was a grand send-off for them all.

8

THE WHEEL OF FORTUNE TURNS ONCE MORE

THE ATMOSPHERE at the OCTU was that of 'Work hard and Play hard', and Bill threw himself into the life of the School. Much of the training he knew already. With his OTC Certificate 'A' and his record in Ceylon he was quite a mature soldier, but these things didn't seem to matter now; it was what the instructors thought of you there! Physical fitness was the watchword; PT in the morning before breakfast and runs in the evening – three miles a day in *boots*. Bill enjoyed the musketry – there was plenty of shooting both small bore and .303. In this, he was well ahead of the rest of the course and was given others to instruct. Then there was advancing in the field under a screen of machine-gun fire. Two instructors would fire in enfilade from each end of a line of cadets walking forward, lifting the line of fire as they came into it. A chastening experience, both for the course and for the instructors.

Bill had heard very little from Betty since the shipwreck. The airmail was still suspended and the mail via the Cape was long and erratic. The few cables he had had seemed to portray an air of hopelessness which worried him greatly. In all his letters he tried to encourage and re-inspire her. He wished he knew with what success. How had she really fared and what effect had her ordeal had upon her? Waiting to hear was a nightmare, but life had to go on.

One of the most important attributes in the Indian Army (or as a British officer attached) was the ability to speak Urdu, the army version of Hindustani. By the time he would be commissioned, he was expected to be able to command the Indian ranks in that language, and to pass an army Urdu exam within a year. Bill was determined to become as fluent in Urdu as he already was in Tamil. In addition to the sessions on Urdu during the course, Bill hired his own private *munshi*, or Indian language teacher, for whom he had to pay himself. Mr Ram Lal was a dignified little man with thin, metal-rimmed spectacles, who was at pains to list the names of all the English sahibs he had taught. (A few of the course members had engaged female *munshies*, but Bill thought that the Urdu acquired might be more vernacular than grammatical!) Bill and Mr Lal spent hours together in the evenings talking, mainly, but not always, in Urdu. In addition to the language, Bill learnt many things about India which later were to prove invaluable to him.

Bill applied for a Commission in the Royal Artillery. He was interviewed, in common with others, by an RA brigadier. Bill did not relish his appearance. There were many more applicants than vacancies and he had virtually no experience of gunnery. However, for his own and Betty's sake it was vital that he should get a British Commission.

The interview was pleasant enough. Bill stuck his chest out and answered all the questions without hesitation. The brigadier asked about his maths.

'Well, I got my London Matriculation, sir,' said Bill.

'Did you take trigonometry?'

'Yes, this was part of the mathematical papers in matric.'

'Why do you want to go into the Gunners . . .?'

Bill met the questions as they came, but those about maths – particularly trig., worried him. This was a subject introduced briefly at Christ's Hospital in his last year, and that was eight years ago!

Poor Bill. It looked as if he was going to need a *munshi* in maths as well as Urdu!

Soon afterwards he was called in by the CO. To his joy, he had been accepted by the Artillery! Together with Richard Newton, another ex-planter, he would be leaving the following week. He was called upon to sign his own confidential report. He need not have worried about it: his conduct had been 'exemplary'; he was a clear thinker and a good critic; handled men well and had shown the greatest keenness in all tactical and weapon training. Map reading was 'very good' and he was expected to pass his elementary Urdu in two to three months.

He wrote to Betty saying that she had better increase the size of his top hat for their wedding from 6⅞ to 7¼!

Just before he left, he received two letters from her written a month after the shipwreck. They had been round the Cape and were forwarded from Ceylon. They were written before she knew he was leaving for Belgaum. He read them with a heavy heart. She seemed to have been affected by the ordeal even more than he feared. She had said that everything – him, Ceylon, her own family and her home – had become something unreal and not part of her. She had seen her doctor who had treated her for continuing shock.

Poor Bill. What could he do? He wrote to her a most compassionate letter saying that he was still very much alive. He was given three days' leave when he left Belgaum. He went down to Bombay and bought a diamond and zircon bracelet which he posted to her, together with a parcel of food for the family. He sent them with all his love, though in his inner self he couldn't help feeling that there were better ways of entering a woman's heart. Betty would know what he meant.

At Deolali Bill found that he was well behind most of the course in basic knowledge. Most of the members had been in the ranks of British regiments before being selected as cadets. Some had come straight from England to the school. On the other hand, Bill had no experience of field gunnery – he did not even know the difference between a piece and a cradle! As he had expected, the maths was 'tricky' to say the least. He had to use what little spare time there was to get extra tuition on bearing-picket procedures and log conversion factors: sin 'Y', cos 'Z' and tan 'S'. Ugh!

Gradually, his efforts began to tell and things began to fall into place. The Instructor of Gunnery, a captain, was quite young, about Bill's own age, albeit a graduate in mathematics at Oxford! However, he was quite approachable. More importantly, Bill learnt from discussion with other course members. There was not much else to talk about; the subject of gunnery became obsessive.

As Bill's confidence began to return and he came to know his fellow course members, so he began to enjoy his training. If only he could hear from Betty. The Overseas mail seemed to have dried up and he doubted whether his letters were getting to her. It was agony not hearing, particularly in view of her letters after the shipwreck. The news was grim too: two battleships had been sunk in the Far East by the Japanese, and there were reports of Allied troops being taken prisoner.

The Airgraph letter service for forces serving overseas had started. One had to write on a special form a one-page message, which was then reduced onto microfilm and reprinted in England. The process was only one way to start with, but at least he could write a letter that she would get in a fortnight. Then, to his joy, a whole batch of her letters arrived, written after he had left Ceylon. She had received his cable that he was going to Belgaum. They were the most loving letters; she had quite recovered! Perhaps it was that he was now in the full-time services or that there was a faint chance that he would return Home. Who knows? The letters thrilled and warmed his heart.

Their final shoot on the course was an exercise to lay down a predicted screen of fire, on calculated data, without ranging, with the object of achieving surprise upon the enemy. It was also their first night occupation. This was the severest test of their training and involved every man doing his task with the utmost accuracy. The target had been identified in daylight the evening before, and located on the maps. A wrong reading would have ruined the whole shoot. The gun positions were pegged for each gun and the line, angle of sight and range calculated. Variations in the condition of ammunition, the wear of the gun barrel and the atmosphere (this latter applied at the last moment) were calculated and incorporated in the lay of each gun.

The night occupation was of itself no easy task, with hooded lights and the minimum of noise. With the guns laid, the course director came to check each gun against the data calculated. Then they waited, tensely, for the dawn. Slowly, very slowly, the first rays of light appeared in the East, then the target could be seen in the early morning haze. 'Prepare to fire,' came the order. 'Five rounds, Troop fire, FIRE.'

All the hours of study, sweat and toil seemed to be embedded in those rounds as they sped towards the target. The noise from the guns, in contrast to the early morning stillness, was deafening. 'Target destroyed,' said the Range Officer. A cheer rose from the troop, as the cadets hugged each other. The twenty rounds had indeed landed in the target area. The instructors were as pleased as the course members.

After that, what more could they be taught about gunnery?

Things moved quickly after that. Bill's name was posted among those to be commissioned. This was by no means exhaustive; two cadets were to be 'returned for Infantry service' and others, including some of those from British regiments, were 'retained for further training'. Bill cabled Betty and had a reply by return: 'Congratulations, darling, so proud, lovingly, Betty.'

Their postings came through amidst a good deal of excitement. Bill was to join the 8th (Sikh) Light Anti-Aircraft Regiment, at present in Peshawar, North-West Frontier.

On the last night before they were commissioned there was a party at the Club at which the celebrations were fast and furious. As midnight approached, they stood on a dais and, as the clock struck, they solemnly pinned on each other's 'pips'. After which, every girl in the room kissed each newly commissioned 'officer'. It was all good fun, but Bill's train for Delhi left at 1 a.m. so he had to dash.

As the train wound its way up through Lahore, Rawalpindi and Nowshera it began to get noticeably colder. The Punjab was in places lush and green, with cultivated fields growing cereals. The people, mostly Sikhs, were taller, stronger-looking and personable with their beards and well-wound *pugris* (turbans). Everyone seemed to be occupied, moving with a sense of purpose and a certain silent dignity. Gone were the rows of people squatting on their hunkers in the streets 'waiting for something to happen'. As they climbed up to Peshawar with the Himalayas behind towering in the distance, Bill felt a thrill running down his spine. After all, he had been born in the Canadian Rockies! This was to be a new experience.

He was met at the station by a sepoy in a smartly wound blue and red *pugri* and an 'I. Arty' on his epaulettes. Saluting smartly:

'Baker Sahib?'

'*Ji han*,' replied Bill.

'*Pas ana Sahib*,' said the sepoy leading the way to his truck.

Bill entered the ante-room of the Adjutant's Office to be met by the Regimental Clerk, Havildar Banta Singh, who addressed him in English. 'Please sit down, sir, the Adjutant will see you shortly.'

Ushered into the Adjutant's office, Bill saluted.

'Welcome, Baker. I'm John Graham,' he said indicating a chair. 'I gather that you came from Ceylon and are a Tamil speaker. How's the Urdu going?'

'Well, Captain, I have been having additional sessions with *munshies* at Belgaum and at Deolali; I'm hoping to take the army exam fairly soon.'

'That's just as well. We have just had a number of English officers from Home who are only just beginning the language. We are in need of Urdu-speaking officers.'

He went on to explain that the regiment had recently been formed by the milking of the 4th/11th Sikh Infantry. Viceroy's commissioned officers, NCOs and sepoys had been supplied, but needed training. There were also some Indian commissioned officers who had come from Deolali. Major Bandari was a Regular, trained at Sandhurst. Bill would be in 'A' Troop under Lt. Gurbitab Singh. The first parade would be at 0630 hrs for PT and gun drill. The CO, Col. Spender, would see him after breakfast tomorrow at 0830 hrs. A bearer was waiting for him and would show him his quarters.

Banta Singh was still at his desk. Looking up: 'This is your bearer, Sahib. He is a Pathan. His name is Sayed.' Bill looked at the man who was to be his servant and who was to wait in Mess. He was tall, over six feet and lithesome. He had a large *pugri* with a 'tail' flowing down his back. He wore the spreading type of Mohammedan trouser. He was capable and seemed to know everyone in the Mess.

Mess that night was informal. There were dinner nights three times a week. Bill began to sort out in his mind who was who. There was a core of regular Indian Army officers, the CO, the Second-in-Command and Major Bandari the Hindu. Then there were two English subalterns from Deolali and four Sikhs. The rest, including a rather flamboyant major – a cockney commissioned from the ranks – had just arrived from England, and were battle experienced.

Untypically, Bill decided to hold his peace in the Mess for a night or so. He was 'on neutral ground' between the factions. The situation would at least be interesting!

Bill was duly summoned the next morning to appear before the CO. A serious-minded man in his early fifties, he was not the blustering type that one sometimes found. He enquired what Bill had been doing. After Bill had spoken; 'Then you are used to handling Indian labour?'

'Yes, sir. That has been my job – but, of course, Tamil labour, not Sikhs.'

'Your reports say that you are good at languages. Have you passed your Urdu?'

'Not yet, sir. I hope to next month.'

'I think that you should know that we have here two batteries of Sikhs and one battery of Punjabi Mussulman Pathans. These are among the best soldiers in the Indian Army. Both Sikhs and Pathans have a long history of service to the Raj, but they have different religions and different backgrounds. Any dispute or unrest in the one will not necessarily be shared by the other. The postings within the regiment are, therefore, intentional. Both races are respecters of fair discipline. I am sure that you know what I mean. If at any time you come across anything of which, from your experience, you think that I should be aware, you are free to approach the Adjutant.'

'Thank you, sir,' said Bill, saluting as he left. What, of course was important was what the Colonel had not said. Bill had read enough of military history and had listened both to Leicester Green and Colonel Gorringe to know that the Mutiny still lingered in the minds of these older soldiers. It was nice to feel that his experience had been recognised.

It was to be a formal Mess Night that evening. Shortly before dinner Sayed appeared. He looked splendid in a spotless white *pugri* and a long white linen coat with regimental buttons. Round his waist was a thick cummerbund in red and gold braid. He asked for the Sahib's approval.

Bill was somewhat aghast by the Mess Night. It was just like a formal dinner night in peacetime. There was the long polished mahogany dining table immaculately laid with the regimental silver, crockery and flatware. Just as Bill had seen at Winchester in 1939! The only difference was that the officers were in khaki field service order. By the CO's decree, it was the night when all conversation had to be in Urdu. Bill was sitting next to Lt. Gurbitab Singh, his troop commander, who was anxious to find out what this new subaltern was made of. Gurbitab spoke clearly and correctly. Soon they were both laughing at the antics of Mr Lal at Belgaum, for whom Bill still had a tender regard. Bill had noticed, out of the corner of his eye, that the recently arrived English officers were sitting in an embarrassed and self-conscious silence, unable to participate. No doubt it had been the CO's intention to spur them on with their studies.

Major Donnor, the cockney, was siting next to Major Gail, the Second-in-Command, and opposite to Major Bhandari. Suddenly Donner stood up and said in a loud voice: 'Does anyone know there's a war on? I came out all this way to teach these natives how to fire guns, not to sit around a table while everyone else is talking a foreign language!' With that he stormed out of the Mess.

It was Major Bhandari who broke the deadly silence that followed. '*Yih yakhni maujud achchha hai*,' (This soup is very good) he said.

It was with relief that conversation resumed. Taffy Williams, one of the English contingent, leaned across the table to whisper: 'The poor man has just lost his wife in the London bombing.' There was great sympathy for him, but emotion was not something that British officers were expected to show before Sikhs and Pathans.

Bill learnt from the Adjutant that the army Urdu exam, which was only held every six months, would be held, for local entrants, in Peshawar in three weeks' time. Did Bill wish to enter? Since Belgaum and his sessions with Mr Lal, Bill had done very little Urdu conversation. He had taken his books with him to Deolali, of course, but that was entirely a British establishment with no time for *munshies*. In fact, those cadets going to British regiments never learnt Urdu. 'Yes,' said Bill, but, explaining the position: 'can I have time off between now and then, please?'

The exam, run by the Army Council, was a formal affair. Text books were supplied to candidates and there would be 'unseen' papers as well. There would be oral and written examinations. Pay would be increased for successful entrants.

Poor Bill – more exams! There was no lack of good *munshies* in Peshawar; after all, it was one of the centres of the Indian Army. Sita Ram was one of the old school who was used to dealing with young officers. Provided you did as he wished, he was an excellent tutor. He and Bill became completely engrossed; by the end of the three weeks he had read and digested the text. He and Sita Ram had spoken nothing but Urdu and in the end the old man complimented Bill on his determination.

The exam was held at Government House in Peshawar. The oral was taken by an Indian Army Sikh major who did not beat about the bush with his conversation pieces. As he left, Bill was told that he had passed.

The written papers were more tricky and took the whole afternoon. The first was on the text which he had just mastered. The Urdu/English and English/Urdu translations were taken almost directly from the book! The 'unseen' ranged over military history, current affairs and colloquial conversation, which old Sita Ram had cunningly introduced during the tutorials – the old devil! But what would Bill have done without him?

Bill had to wait a week before he heard that he had passed the Elementary Urdu. There was not much recognition in the regiment, although the CO called him in. He was the first new entrant in his wartime command to pass. The results were gazetted in the Indian Press. Bickford, who came over with Bill's contingent, was the only one from that group to pass as well. He had been over for five and a half months of which two were at Deolali.

The only trumpet-blowing that he could do was to Betty in his next letter!

At the end of May 1942 reports were coming in that there had been air raids on China Bay and Colombo in Ceylon. The news was heavily censored, but it seemed that the Japanese aircraft had been beaten off by the RAF and the Fleet Air Arm which had destroyed six bombers or fighters with six more probably accounted for. Allied casualties were not quoted although 'some of our aircraft are missing'. The episode was put out more as a nuisance than of military significance. On the nuisance side it seemed that the life of Colombo had been brought to a halt with shops closed, banking facilities suspended, and sweepers (the *varsi-kutties* who clear the toilets) non-existent.

On the more serious side, it seemed that the Japanese Fleet had penetrated the Indian Ocean and the Bay of Bengal where they had been sighted off Vizagapatam. There was speculation about what would happen next, and whether this was a prelude to more serious aggression on the part of the Japanese.

To Bill this seemed rather unreal. He had plenty of other things to occupy his mind; this attack on his previous homeland was not now his concern – except, of course, that many of his friends were still there. But how different it might have been. Suppose that Betty had managed to get out to him there and that he had had to leave her behind on his drafting to Belgaum . . .?

Although the main duty of Bill's newly formed regiment in Peshawar was to train in anti-aircraft gunnery, it also had a role, in conjunction with other regiments on the station, to monitor and keep open the trade route between India and Afghanistan, by way of the Khyber Pass. This wild track of Himalayan country was a fruitful ground for marauding tribesmen, mainly of Afghan origin, who were apt to plunder the colourful caravans of camels plying their trade. Not only was there a steady stream of silk, wool, carpets and hangings coming southwards into India, but in return, rice, wheat, leather, silver and arms were making their way north. In addition, rivalries in the past between different tribes had created much bitterness; it was not a happy region.

High up the Pass, near the col into Afghanistan, there was a cantonment occupied by troops on policing duties. At an altitude of some seven thousand feet it was in a commanding position looking down upon the winding road below, but it was a desolate place.

Not long after Bill had passed his exam, 'A' Troop was ordered to occupy the cantonment at Landikotal for six weeks. Gurbitab Singh had been briefed that there had recently been some rather disturbing incidents in the area. He was to keep both his CO and Government House advised. It had been the hot season in Peshawar – how lucky they were to be going to a hill station!

The 'A' Troop convoy consisted of Gurbitab and Bill with two jemadars (Viceroy's commissioned officers), four havildars, forty sepoys and ten followers. Lined up on the tarmac were their two 15-cwt trucks, and four 'Spiders' carrying the gun crews and towing the gun carriages and Bofors guns. Three 3-ton trucks carried the stores and followers. The CO took the salute as they moved off headed by Lt. Gurbitab Singh. Bill's 15-cwt with one of the jemadars brought up the rear. All personnel were armed.

Bill was determined to enjoy this, his first visit up to Landikotal. As they drove higher and into the heart of the mountains the scenery became increasingly majestic. Vast pillars of rock reached up to the sky from the greenery of the valleys below. Every now and then a shepherd, with his goats, would stand watching them. The kids, taking advantage of the pause, would suckle their mums. Bill had a strange feeling that this was home to him; silly of him, of course. It was just that God's hand on nature became real and tangible.

The road became increasingly exacting on the young Sikh drivers, many of whom looked just lads. For some this was their first experience of the Pass.

Suddenly the convoy came to a halt. Drumming his fingers on the side of the truck, Bill sent a despatch rider on to find out why.

The sepoy returned looking distressed. '*Jaldi, Jaldi, Lt. Sahib. Haldisa maujud hota hai.*'

Bill grabbed the man's motorbike and tore up the road to the scene of the accident. The havildar and gun crew of one of the Spiders was parked precariously near the outer edge of the road. There was no sign of the gun, or the gun carriage. It seemed that in turning a sharp bend, the engine draft connector between the Spider and the carriage had sheared away. Both the carriage and the gun had fallen into the ravine some one thousand five hundred feet below. The driver of the vehicle was clearly upset and Bill comforted him. Indeed, the whole crew were lucky not to have been pulled over. By now, Bill's truck had arrived so he drove on to tell Gurbitab. They agreed that nothing could be done about it, except that, if a guide could be found, a working party should be sent to blow up the ammunition. Bill would have to draft the accident report!

They had been at Landikotal for ten days. In view of the reports that had been received, they had been patrolling the Pass at irregular intervals, but had seen nothing awry. On one occasion when Bill had been with the patrol, they were waiting on the Afghanistan border when a caravan of camels stopped to adjust the packing of the load of carpets. Bill walked over to inspect them. They were some of the most beautiful that he had ever seen. The intricate patterns were in delicately washed colours and of exquisite knotting and workmanship. He enquired the prices. 'This one Sahib Rs. 200, that one Rs. 250, that small one Rs. 100.' If only Bill had a home to put them in!

On returning to the cantonment, Bill was met by Gurbitab holding a message labelled IMMEDIATE AND CONFIDENTIAL. It was from Regimental HQ.

It seemed that, following constant plundering of villages in the valley of the river Swat north of Peshawar, the police had caught one of the gangs, who were being held in Peshawar jail. Two were on charges of murder. Another gang of Afghan tribesmen numbering about ten men had raided the house of the Headman for the area and abducted his wife and young son. The gang was armed and had been sighted in the mountains. It was thought that they were making for Jalalabad across the border. They had left a note in one of the villages that the lives of the hostages would be taken if the prisoners were not released.

Government House took the most serious view of the incident; the principles of 'Local Rule' were at stake. 'A' Troop would be joined by representatives of the police and two trackers to act as guides.

'A' Troop was ordered to take all military measures necessary to recover the hostages alive.

'This is a job for you,' said Gurbitab. 'Let me know in half an hour what men and equipment you want.'

Two police sergeants and the two trackers had arrived while Bill was getting his orders. After discussions with them, Bill decided to take Jemadar Printab Singh, two havildars and ten sepoys. He would need ammunition for his .45 and that of the Jemadar. The men were to carry forty rounds each. He would need stores and provisions for four days to be transported, with two followers, to a given map reference, described by the guides, where a base camp would be established. Their intention was to intercept the gang before they reached the border. He suggested to the troop commander that a picket be established there as well, although the guides thought that this method of exit to be unlikely.

The detachment departed in the two 15-cwt trucks following the road down the pass that, by now, they knew quite well. Later, the guide indicated a narrow forestry track that Bill had not noticed before, and which led down into the valley below. There they found a wider unmade road which the guides described as being the old drovers' track, used before the main road was built. The track was passable for vehicles up to the foothills, where there was a village; thereafter, the way up to the border was on foot or by mule.

They reached the village just before nightfall. Taking Printab Singh with him, together with a police sergeant, Bill called upon the headman. Yes, a party of Afghans, a rough-looking lot, had passed through that afternoon. He had not seen them, but a woodman returning from the head of the valley had said that he had passed a camp where there was a woman and a child.

He thought that there were about five men and mules – one of which, he remembered, was unusually dappled with a wall eye.

When the mission was explained to him, the headman agreed to help in any way he could. The woodman would accompany them up to the site of the camp. They would leave before first light at 4 a.m. On their return to the vehicles, the cooks had hot soup to go with their hard rations. Lying down under a rock Bill dozed fitfully. He had a plan for the morrow.

It was still dark when they left, but the guides seemed to be able to see in the dark. He described his plan to the Jemadar. The guide and the woodman were to go about three hundred yards ahead and wait until first light to make sure that the Afghans were still there. They would then return to the main party which would then split. Bill would take the woodman guide, the Havildar and eight sepoys up the mountain and to the back of the camp, leaving the Jemadar, the other Havildar and two sepoys on the road. When Bill's party was hidden in position, the woodman would walk openly down the track to the Jemadar. Once Printab Singh knew that the main party was in position he would start firing at the camp making sure not to aim at any of the occupiers, for fear of injuring the hostages. He would then call upon them to surrender. On the assumption that the robbers would flee up the mountain, Bill would be waiting for them . . .

It was still dark when they reached the camp. The woodman and the guide had returned, the latter having identified the dappled mule in the line of ties. The next manoeuvre was the tricky one. As luck would have it, in the first light of the dawn they found a sheepwalk leading above the camp.

The plan worked perfectly. On the first enfilade from Printab Singh the Afghans grabbed the hostages and made off on foot up the mountain into the arms of those waiting for them. One raised a gun at the Havildar and was shot. Another lunged at the woman with a knife, only to receive a rifle butt to the head. The remaining four surrendered. Bill kept them covered as they were disarmed and gleefully handcuffed by the police sergeant. It had all happened so quickly that everyone was a little dazed.

Bill's immediate concern was for the headman's wife and her son. She had not eaten since she was abducted, refusing to take food at the hand of the Afghans. Both she and the small son, aged about six, were exhausted and very distressed.

Hearing the shots, Printab Singh and the second Havildar arrived up and were both surprised, and relieved, that it was all over. It was decided that Bill and the Jemadar together with the sepoys would take the ex-hostages and the prisoners down to the base camp. Bill 'requisitioned' the mules for the woman and her son. He assumed that there would have to be some sort of inquiry over the two dead men. He left the policeman and a havildar at the site and would decide with the village headman what to do with the bodies.

There was elation when they returned to the base camp. Arrack was produced and they drank to the success of their mission. But Bill was anxious to get the woman to hospital. Although thankful to have been saved she felt defiled. She would only take a little food. He made a bed for her on the floor of the truck, and taking a havildar and an extra sepoy with him ordered the driver to return with him to Peshawar. The headman would look after the police sergeant whom he had left up at the site. The Jemadar was to take charge of the prisoners and return them to the cantonment awaiting the arrival of the police.

It was dark by the time Bill returned to Landikotal. He reported to Gurbitab who seemed a little put out that Bill had not reported to him before taking the hostages down to Peshawar, until Bill explained the need to get the woman to hospital. The prisoners had arrived at the camp and were now in the charge of the two police sergeants. He hoped that they would soon be moved. If only he had a telephone! Bill excused himself – he was dying for a bath. The arrangements were primitive, but a follower brought a hip bath into his cubicle and two buckets of warm water. Then he knocked up Printab Singh, the Jemadar, asking him to take him round the quarters of those who had been out with him. He spoke for some time with the havildar who had shot the tribesman; he seemed quite imperturbable, Bill was thankful to see. The sepoy who had prevented the woman being stabbed also seemed to have no ill after-effects. Bill told them all how pleased the 'Government Sahib' had been.

The cooks produced a good curry that night, after which Bill retired to his cubicle to read his letters.

The letter from Betty had taken seven weeks to come out. At the end of 1942 the war at Home was at its height, with continual night bombing and daylight dogfights in the skies above. Betty was an ARP warden at night and working for her father during the day. It was incredible that she could find the time to write such loving letters to him. Their lives seemed so different, and Bill felt so out of things.

The Airgraph was written only ten days ago. It was only a short one-page photocopy of her handwriting; it brought her closer to him, but oh! she had a threatened appendicitis. Here he was, miles from anywhere, with no way of finding out how she was. What a funny old world it was.

Tired now, he lay on his palliasse contemplating the day. He was not without pangs of conscience himself about the death of the two tribesmen. Still, they were all armed – it might have been much worse . . .

Early the next day two police vans and an escort arrived to take the prisoners and to fetch the two police sergeants. Bill called the sergeants to one side before they left, to thank them for their help. Then Bill added, 'We were indebted to the woodman for finding the tribesmen's camp. Do you know his name?'

'Ramzan, Sahib,' was the reply.

'Well, do you think that the police could arrange for some sort of commendation to be given to Ramzan?'

'Yes, Sahib.'

Bill watched the prisoners as, still in handcuffs, they were herded into the vans. They were a motley and bedraggled-looking lot, with dirty, dishevelled headcloths above their long, flowing, unkempt beards. The Sergeants were also loading, into the front of the vans, the arms that had been recovered. There was a modern Lee Enfield .303 rifle which had been in the hand of the man, probably the leader, who had been despatched. Then there were some Russian-made rifles and a very antiquated muzzle-loading blunderbuss, which Printab Singh surmised would have exploded if it had been fired!

Gurbitab Singh heaved an obvious sigh of relief once the police and the prisoners had departed. The normal routine of the day had been disturbed. But his relief was short-lived. A despatch rider arrived with the regimental daily Orders, the mail, and a message that the CO was arriving for lunch and would inspect the troop at 1400 hrs.

It was not long before the CO's station-wagon, followed by a 15-cwt truck, was seen coming up the pass. At least one had a good forewarning of visitors.

Colonel Spender was in a good mood. He had had a special commendation from Government House which he wished to pass on to the troop. He also wanted to know whether the rations and quarters were satisfactory. The Colonel did not say anything specifically to Bill, but several times he felt as if he was under scrutiny. Afterwards, with the troop paraded before him, the CO thanked everyone for the part they had played in the episode. He did not want to mention names; the whole troop had upheld the good tradition of the Indian Army. He was proud to pass on the commendations from the Government Agent and the thanks of the local police.

THE WHEEL OF FORTUNE TURNS ONCE MORE   173

Calling the troop to attention, Lt. Gurbitab Singh saluted the CO as he left. Then a fatigue was detailed to unload stores from the 15-cwt truck.

'A' Troop's period of duty at Landikotal had come to an end. Most were glad to return to Peshawar with its off-duty entertainments. On the day after their return, the CO called Bill into his office, showing him a seat.

'I gather that you did quite well up at the Camp and that the recovery of the hostages was largely of your doing.'

'Well, not really, sir, it was a joint effort by everyone. The Sikhs were so enthusiastic – it was a pleasure to serve with them.'

'It was really that which I wanted to talk to you about,' said the Colonel. 'You obviously have advantages over the other European officers in your knowledge of the East, and you are picking up the language here very quickly. Besides, the Indian ranks seem to like you, though that is not always a good thing if it is overdone. I want you to be the Assistant Adjutant to Captain Graham and to act as Regimental Intelligence Officer. You will not, of course, have disciplinary powers, but I want you to know what is going on and keep me in touch. I expect that Captain Graham will have a number of commitments for you!'

'Thank you, sir,' said Bill. 'I will do my best.'

On the way out Bill saw John Graham. He was to have a desk in the Adjutant's office and to deputise for him when he was out. He was also to be the Motorised Training Officer, and, horror of horrors, Officers' Mess Secretary. Graham saw the look on Bill's face. 'Yes, I know,' the Adjutant said. 'There will be plenty of time for you to be Intelligence Officer as well!'

As he walked out of the Regimental Office, Bill wasn't quite sure what had hit him!

The next evening, when Bill had had his bath and was almost ready for dinner, there was a knock at the door. It was Sita Ram.

'Salaam, Sahib,' said the old man.

'Salaam Sita, *pas ana*,' replied Bill, indicating a chair.

'I have heard about the Sahib's *Bara Ma'rke ka kam* (exploits) at Landikotal. Now I hear that the Colonel Sahib has given Baker Sahib a new appointment.'

'You old devil! How do you know all this?'

'Oh, people talk, Sahib.'

Sensing that his old tutor had not just come to talk: 'Well, what can I do for you, Sita?' enquired Bill.

'Sahib will now want to take his Higher Urdu army exam? Sahib will have very good prospects in Army.'

'Oh, Sita, but I have only just completed the Elementary. Why so soon?'

'Better now than later, Sahib.' Then, continuing without a reply: 'Will Sahib want to take the Persian or Nagri script?'

'Oh, yes, if you say so, Sita. I will take the Nagri. But now I must go into the Mess or I shall be late.'

As Bill walked from his room to the Mess he wondered how the old man knew so much about him; he supposed that a certain amount of backchat about the officers was inevitable, but this was a bit much. Indeed, had the CO anything to do with Sita Ram? He hoped not!

Bill's new appointment opened up a new dimension. One of his first tasks was to get to know all the Viceroy's Commissioned Officers, starting with the Regimental Subahdar Major. He was a huge man towering to six feet two inches by the name of Ujhaga Singh, with a voice like a rhinoceros. His mere presence would put the fear of God into the heart of a young sepoy, or, for that matter, a young British officer as well! Bill felt that he would have to 'tame' him. Then there were subahdars to each of the batteries and a jemadar to each troop. Bill got to know all their names and found opportunities to talk to all of them. Bill found them ready enough to talk – about themselves, their families, their villages and about those in their command. He found the Pathans more reticent than the Sikhs; they felt that they were attached to the regiment. They were in the minority and it was not their own unit. Perhaps they thought more deeply than the Sikhs. There was absolutely no doubt that both the Sikhs and the Pathans were fiercely proud of their race and creed, were of magnificent physique and were fine soldiers. For his part, Bill was also proud to serve with them.

Sita Ram lost no time in the initiation of his pupil. Miraculously, he already had the books (second-hand at a small price) and plenty of time at his disposal to fit in with the Baker Sahib's revered requirements.

But somehow the old man's intelligence had gone awry. The regiment was to move to Vizagapatam on the Bay of Bengal.

The CO called the Battery Commanders into his office, together with the Second-in-Command. The Adjutant, the Quartermaster and Bill were also to be present.

The CO's orders had been carefully prepared, but they were only an outline of the final Regimental Orders. The move would be in three weeks time. Those present at the briefing would fly down to Vizag in two days' time to reconnoitre the positions for RHQ, the batteries and gun positions. The Adjutant and Battery Commanders were to have draft movement orders ready in ten days time. There was an air of suppressed excitement when the meeting was over.

The Quartermaster, Lt. Ranjit Singh, and Bill were the only subalterns on the reconnaissance party. As Assistant Adjutant, Bill amongst his other duties was given the job to prepare a regimental map (to be kept under lock at RHQ and Bty HQ) showing the positions, with map references, for the deployment of all units in the regiment. They were billeted, for their visit, on the RA Coast Defence Battery who were grand fellows, but their hospitality was stretched to the limit. Moreover they provided transport!

For the next ten days the Adjutant's office was in turmoil. Finally, the movement orders were drafted and approved by the CO. The three-day journey to Vizag would be by rail. They would require two trains: one for the officers, VCOs and Indian other ranks, and the second for the vehicles, gun-towing vehicles (spiders), guns, gun carriages and ammunition. The latter would be accompanied by the Quartermaster, drivers, and gun fitters together with a subahdar and jemadar.

The departure from Peshawar was an emotional affair. Many, including some of the officers, had formed attachments, or had families living on Station. 'Living out' was not allowed at Vizag. Even Sita Ram came to see Bill off, so he thrust a Rs. 10 note into the old man's hand. Conditions on the troop train were cramped, the ventilation was poor and the toilets soon indescribable. All meals had to be taken off the train which meant long stops. During these the subahdars organised physical training in which the officers joined. Everyone was glad when they reached Vizag and the deployment could proceed. Although not a major port, the town lies about halfway between Madras in the south and Calcutta in the north. Indeed, when the Japanese fleet penetrated the Indian Ocean for their air-raids on Colombo and China Bay, the intelligence was that a raid might be made on Vizag as well. Now, with the advance of Japanese troops on the other side of the Bay of Bengal into Burma, a landing on the Indian mainland again seemed possible. The troop positions around the harbour had been sited for both anti-aircraft and sea-based targets; the fields of fire for the latter were now widely dispersed, each a unit on its own – gone the days of regimental dinner nights and uniformed bearers!

When the guns were in position Bill was ordered to arrange some 'live' practice laying for them. He was able to negotiate through the local flying club for Mr Sher Khan to fly the club's Tiger Moth at a charge of Rs. 100 per hour. As Bill had drafted the map of the gun positions, he flew with the pilot to show him – it was quite an experience and fine practice for the crews. 'Book him again!' said the Colonel.

The CO had been invited to dine with Colonel Robinson, the officer-in-charge of the Coast Defence Regiment. It seemed that Colonel Spender had been extolling the prowess of his own gunners in tracking the Tiger Moth. Not to be outdone, Robinson described his own tracking. He had hired an old fishing trawler owned by Haidar Ali, who lived in the port. The arrangement was that Ali had been supplied with a hawser eight hundred and eighty yards long which towed a raft of empty oil drums carrying a wooden superstructure. Within a carefully determined range, the six-inch guns were allowed to fire live ammunition at the raft. Ali had always been nervous of the arrangement; he had insisted that firing should not start until he had indicated by a flare that he was ready. Colonel Robinson explained that the object for his gunners had been to bracket, rather than to destroy the raft.

Colonel Spender came back to RHQ full of the idea. Why shouldn't Mr Haidar Ali be hired for his own guns to practice in a sea target role? Captain Graham was to invite Mr Ali to attend the CO's office. Ali duly arrived and disappeared into the CO's presence. Through the door, John and Bill heard some rather heated discussion. After Ali had left nothing further was mentioned, and Graham thought it tactful not to enquire.

It so happened that, at about this time, two of the troop commanders had reported trouble with the prediction systems on certain of their guns. Once laid upon the target, the 584 Predictor calculated the line, range, angle of sight and rate of change resulting from the speed of the target before supplying the data to the Bofors gun by electronic cable fed from a powerful generator. It was of the essence that, when setting up the equipment, both the gun and predictor should be laid in vertical and horizontal symmetry. The trouble had arisen because, for some reason, the two guns had, during practice, suddenly jumped exactly twenty degrees out of horizontal alignment. The gun fitters had been working ceaselessly and had reported the fault rectified.

Meanwhile Mr Ali, who had been demanding far more for his services than the CO had authority to pay, was summoned again. Army command had decided to pay. All was now sweetness and light. The practice would take place the following week. The CO and the other battery commanders would view the practice. An invitation was also extended to Colonel Robinson. Two troops would fire on the first day; it was intended that the others would follow later. On a signal from the shore Ali, at the wheel of the trawler, slowly turned across the mouth of the harbour towing the target behind him. Once this had straightened, Ali shot the flare. 'C' troop was the first to fire. No. 1 gun neatly straddled the target with bursts before and behind the target. Then No. 2 gun followed with a burst before. A single round hit the sea behind the target, then, horror of horrors, the gun jumped out of phase. 'Stop firing!' yelled the Range Officer, and the Gun Position Officer, simultaneously. But not before three rounds had gone – two over the deck of the trawler and one slap into the superstructure. In the next thirty seconds, Mr Ali expended his entire stock of flares.

On shore, all the Colonel Spender could find to say was, 'Oh dear.' Colonel Robinson looked at his watch. 'Well, I really must be getting . . .'
'Yes, of course,' said Spender.

Later, it transpired that the twenty degree 'jump' on the gun was the exact amount of deflection required to turn the gun onto the trawler. A cruel fate! The incident rumbled around RHQ for weeks. Eventually, a naval architect came up from China Bay to assess the damage. Significantly, Mr Ali was able to buy himself a much better trawler!

11 June 1943 was a Sunday. After church parade, Bill retired to his cubicle with his thoughts and to write his weekly letter to Betty. It was now four years since that fateful evening when he had asked her to marry him. Since then they had suffered the agonies of separation; the two abortive attempts for her to join him in Ceylon and the final shipwreck followed by her period, now thankfully overcome, of acute depression. Then there were his own failed attempts to return to England and the now seemingly endless waiting for something to happen. Betty's loyalty to him was incredible. The war was all around her. Nearly all her friends were on active service; a cousin had just lost his life in the RAF. The girls were either married and having babies, or in the services. What could she tell them of her fiancé? That he was stuck in this hole? Bill had now been in a tropical climate for ten years, with only four months in England during the whole of that time. He had actually been mobilized with the Ceylon Defence Forces since September 1938. Surely he was entitled to repatriation, or at least a Home leave? He would try again. But there was no good in raising Betty's hopes. He must reply to her letters and Airgraph. What could he write that would pass the censor? That he

loved her with all his heart, that he longed for her, that she was never out of his thoughts and that he had dreamt how she would look on their wedding day. That he admired her courage and fortitude and above all her loyalty to him. But how utterly inadequate these sayings were; and what hope could he give her for the future?

Bill drew himself up with a start. He was being self-pitying and dramatic. He must write courageously and with confidence. A man without hope is craven indeed. After he had finished his letter to Betty, he wrote yet another formal request for repatriation to England and to serve on the European front, giving his length of service abroad and his domestic reasons. Later, Colonel Spender called him in. He had every sympathy for him and would forward his request to Army Command, but he did not hold out much hope. The emphasis of the war seemed to be passing to the East. With his experience, Bill had a valuable contribution to make to the war effort. Trying hard not to be dispirited, Bill returned to his work.

Some time later, Bill was returning to his cubicle to get ready for dinner. He had spent the afternoon checking wine lists and the menus in the Mess for the following week. He was looking forward to his bath. On his way he was met by Subahdar Major Ugahar Singh. 'Salaam, Baker Sahib,' he said.

'Salaam, Subahdar Major,' Bill replied. What on earth could the great man want with him at this time in the evening?

'Sahib, *bahut taklif Chhaoni men hota hai. Tamil log, Sikh khaly, bahut janjal, janjal majud hota. Sahib ana, miharbani, jaldi ana!*' Poor Ugahar, he could cope with any Jat or Sikh, but these Tamils – he couldn't understand a word of what they were saying!

The regiment employed a number of Southern Indians as followers. These were storemen, clerks, writers, cloth cutters, dhobies, cleaners and other trades. Some had set up small shops, buying goods in the town and reselling them in the camp. Most of the southern races spoke, or could understand, Tamil. The Sikhs, regarding themselves as superior, had little desire to learn. As Ugahar led him, Bill could hear the sounds of strife. Voices were being raised in anger; sticks were waving aloft. Dark now, torch flares shone upon belligerent and threatening faces. It was not a pretty sight. Bill jumped up on the bonnet of a nearby truck, beckoning Ugahar Singh to follow him.

'*Tamil al elam nikka. Unude Vai Mudu. Pesathe! Sikh log band karo. Multawi ek dum. Chup.*'

The Tamils were so surprised at being spoken to in their own language that there was silence. The Sikhs followed suit. Bill followed in both tongues, separately.

'I have no idea what this trouble is about, but your behaviour is disgraceful. I cannot talk to you all. You are to appoint two persons each to represent you. When we have found out from them the cause of the complaint, we will reach a fair settlement. Now, tell the Subahdar Major whom you want to speak for you.'

Ugahar Singh found an empty hut with a table and chairs. As usual with such disputes, the issue was a simple one. Some of the Sikhs were hooked upon opium. The attitude of the regiment to the habit was, officially, to ban the use of it. But it was well known that sepoys, even on recruitment, were users. Provided it was taken in moderation, a blind eye was turned. However, when efficiency was affected, addicts were expelled from the Army. What had happened now was that the Madrasses had set up boutiques in the camp to sell 'doctored' opium at a high price. Those creating the disturbances were, to some extent 'under the influence'.

Bill was late for dinner and incurred the displeasure of the Second-in-Command.

'Why are you late, Baker?

'I'm sorry, sir. I had a job down in the lines.'

'What have you been doing?' persisted the Major.

'Well, sir, there was a bit of a guffle between the Tamils and the Sikhs. Ugahar Singh asked me to go down.'

'Oh, those bloody Tamils,' said the Major.

'Yes sir,' replied Bill.

Next day, after discussion with John Graham, Bill went down to the government analyst in Vizag. He was most helpful. He would make visits to see that only pure opium was sold, and that the price was fair. Nothing more was said about the matter. The CO had said, more than once, that the Sikhs are fine fellows, provided they are required to work hard and play hard. It's when they have nothing to do that . . .

There had been some grizzling in 'C' Troop. 'I think that the men want a change of face and scenery for a bit. I want you to take a party inland for a week with hard rations. The maps here are quite good, but we want more detail. Fill in what you find on this area,' the CO said, pointing to the map. The was just like the days at Trinco, but the countryside was kinder and more productive. Bill set off with a three-ton truck and two 15-cwt vehicles, a jemadar, two havildars, and twenty men. He was to keep them fully occupied. Leaving the vehicles in the hinterland in the charge of a picket, they split into parties to follow the rivers and paths, the forests and fields, mapping as they went. Bill discovered a plantation of cinchona trees, from the bark of which quinine is made. There was a dire need of supplies at the time. He would let Army HQ know when he got back. On another occasion, when they had been out all day in the heat, they entered a field of pineapple, just ripe to perfection. The sepoys fell upon them. Bill hadn't the heart to stop them – and probably couldn't if he had tried. Drawing their knives they sliced the fruit off the plant, decapitated it and sank their teeth into the juice. The jemadar handed one to Bill – it was nectar. They left before the felony was discovered.

In the evenings, they sat around a camp fire talking. They would tell Bill about their villages, their families, and, embarrassingly, what they thought of their officers. Bill did not stop them. He learnt more Urdu that week than in a month of *munshies*, and about the regiment too, but he would forget that. He was more interested in finding out how the Sikh mind worked than in a little tittle tattle.

They were back at RHQ. The CO had called Bill in. Regrettably, there was to be a court martial of a sepoy on the grounds of mutiny, the penalty for which could be death, if found guilty. The case was to be heard before a panel of the Judge Advocate General's Department, who would come up from Madras. Lt. Pertaub Singh would be the Prosecuting Officer. The CO wanted Bill to be the Defending Officer; Bill caught his breath. The law was hardly his strong subject – would that he had his friend Noel Gratien in his pocket. 'Yes, sir, if you wish me to.' The CO nodded. Bill was given a copy of the charge sheet and a copy of the proof of evidence for the prosecution. He set off to see the prisoner, who was being held in the jail in Vizag, there being no adequate facility within the regiment.

Bill found Ram Singh sitting with his *pugri* removed and bare-legged upon a stool in the corner of his cell. He was deep in thought, but there was an air of defiance and dignity about the man. He looked up when Bill was let into the cell by a warder, but did not speak. Bill sat down on the only chair and put his papers on the table. 'I have come to help you,' he said.

The silence seemed endless while Ram summed up his visitor. Eventually: 'I have nothing to be ashamed of,' he said.

'Of course not, but do you feel like telling me about it?'

'Why should I, Sahib; will it do me any good?'

'Under British justice, you have a right to be represented in court by an officer. I shall do everything I can to see that you get a fair trial.'

There was a further silence. Then, Ram Singh got up, put on his *pugri*, and stood before his Defending Officer like the soldier he was.

'Sahib, I am a Rajput Sikh. Havildar Govind Singh, who is my accuser, is a Jat. In history, we Rajputs are a superior race, but many years ago we refused to join the Jat in the revolt against the Mogul Empire. Since then we have been overshadowed, persecuted and taunted by the Jats. It is my misfortune to have been recruited into a mainly Jat regiment. I was young when I joined and did not appreciate the significance. Things have not been easy between the Havildar and myself; he has continually taken a superior social attitude to me and has taunted me about my religion. Finally he put around a story that I had cut my hair, which is untrue. On the day in question, he uttered a foul word at me; I was so incensed that I raised my rifle and shook it at him. He took it from me and I have not seen it since. He reported me to the Jemadar, who arrested me.'

Bill looked at the prisoner. The story had been delivered in a spontaneous and dignified manner – he did not think that it had been made up.

'Ram Singh, at this stage I have three questions for you; please answer them truthfully. Firstly, have you ever tried to get any other sepoy to join you in any form of disobedience to the Havildar or any other NCO?'

'No, Sahib, why should I? This is a personal matter between myself and the Havildar. I have no complaint against the Army.'

'Did you aim the rifle at the Havildar?' said Bill.

'No, Sahib, I shook it.'

'When you shook the rifle, was it loaded?'

'No, Sahib, ammunition is not normally on issue to sepoys. It is kept at Troop HQ and issued for practice or emergency. It had not been issued at the time.'

'Thank you, Ram Singh. I don't think that you have much cause to worry. I will see you again before the trial.'

When Bill returned to RHQ he spoke to Banta Singh, the chief clerk. 'Banta,' said Bill, 'can I please see a copy of the Indian Army Act?'

'Yes, Sahib,' said Banta Singh, getting up from his desk before going to a locked cabinet.

'As a matter of interest, who drafted the mutiny charge on Ram Singh?' said Bill.

'Lieutenant Pertaub Singh, Sahib.'

'And to your knowledge, did he consult the Army Act before so doing?'
'Not to my knowledge.'
'And it is kept under lock and key?'
'Yes, Sahib.'
'And you have the key?'
'Yes, Sahib.'

While he was waiting for the day of the trial, Bill heard that his application for repatriation to England had been rejected outright. He could not be spared from his unit. No reasons were given, and there was no acknowledgement of the time that he had spent Overseas. He stifled his sense of frustration. It was clear that, since the shipwreck, Betty had given up any attempt to join him; indeed, it was not right that she should do so. The waiting, without being able to do anything about it, was unbearable. Bill was also beginning to feel that the war was passing him by. While the military life around him was real enough, the chance of the Japanese either mounting a raid or invading Vizag seemed to be increasingly remote. He was also bothered that there was so little real intelligence about the progress of the war in Burma. Looking through District Command Orders, he saw that officers were required to take command of landing craft on the Burmese coast with the implication that troops would be deposited further south in the Bay of Bengal.

Without any very special qualification, he applied for the job.

The day of the court martial arrived. Those sitting were an Indian from Army command whom Bill took to be a Hindu, a British Major from another regiment and a woman who was a Parsee from the Indian Women's Auxiliary Corps. The charge of Mutiny while on Active Service was read and the Appearances taken. After announcing himself, Bill immediately entered a plea that there was no case to answer to the charge of Mutiny. Such a charge was, by definition, one of incitement to others to disobey. He quoted cases where Mutiny was based on rebellion. There was no evidence of rebellion in the charge, which should therefore be dropped. The Chairman, looking a little nonplussed, said that the Panel would withdraw to consider the plea. After about ten minutes, they returned. The plea had been noted, and would be taken into account. Nevertheless the trial would continue.

In his opening address, Pertaub Singh described the events which led to the incident. It had been Ram Singh's duty to clean the gun sights and surroundings. He had been lazy. When Havildar Govind Singh rebuked Ram Singh, the sepoy had aimed the rifle at him. He thought that he was going to be shot. He took the rifle from the prisoner and confiscated it. He found that the rifle was loaded. The Havildar then corroborated the statement.

Both Pertaub and Govind had spoken in Urdu very quickly; it was not the first time that Bill had to have all his wits about him. Now, in cross-examination, he spoke slowly and deliberately. In what way had Ram Singh been lazy? What words had Govind actually said to Ram? Why should he have felt so aggrieved? What evidence was there to confirm that Ram had pointed the gun? None of Govind's answers satisfied Bill. Then there was the key point about the ammunition. It was the Havildar's own responsibility to issue the ammunition, but he had not done so. It was not until later that he opened the bolt and found the rifle loaded. He did not do so in front of the accused. Bill put it to him that he was lying.

During the trial, Pertaub Singh called the Troop Commander, Captain Williams, who was, of course, a companion in the Mess. Bill sat silently listening to his evidence. When called upon the examine the witness, Bill stood up.

'Captain Williams, I have listened carefully to what you have said. You must remember that this is a serious trial in which a man's life might be at stake. You were not present when this incident occurred; you were not even on the gunsite at the time. Nothing in your statement amounts to first-hand evidence.' Turning to the Panel: 'I ask you, sir, to disregard the evidence of this witness.'

Bill's summing-up was short and sweet. This was a private feud between a sepoy and a Havildar. No evidence of incitement against the Army had been produced. The charge of Mutiny ought never to have been brought. While the prisoner accepts that he was provoked by the manner of his rebuke for his alleged 'laziness', the rebuke was in language not permitted in the Army. There were no witnesses to what then happened. It is one man's word against another. What is certain is that the prisoner did not have access to ammunition for his rifle at the time. The weapon was not unloaded in his presence; he did not know until the next day that he was being charged with aiming a loaded rifle. I submit that the evidence does not justify a conviction.'

The Panel retired. On their return, the prisoner was sentenced to four months' detention. He was to be transferred to another regiment.

In the Mess that evening Captain Williams came up to Bill. 'You rascal,' he said with a smile, 'I'll have you for insubordination.'

Bill held out the palms of his hands. 'Which hand would you like to slap, sir: one, or perhaps both?' They went off for a drink. Pertaub Singh joined them.

Not long after the court martial, the CO told Bill that his name had been put on a list of potential Adjutants for appointment as and when a vacancy occurred. He was pleased that the Colonel had done so, but, at the moment, it was just another uncertainty. Apart from a formal acknowledgement, he had heard nothing more about the Burma landing craft. Then out of the blue came his posting. He was to report to GHQ at Karachi for 'special duties'. He was not told what these duties were, and the Colonel didn't know either. He left Vizag with his kit on a 'blind date'; rumour had it that he would get his captaincy. It was a bumpy flight in a freight aircraft, but it gave him an opportunity to write to Betty. His letter could not conceal an element of excitement about the future.

What were these 'special duties', and where would they take him? As he mounted the steps of GHQ in Karachi, he was soon to learn. The Indian artillery had expended rapidly in 1942 and the early part of 1943. 'It is important that the units are complemented by the right race religion and caste of sepoys and NCOs, and as quickly as possible. Your duty will be, firstly, to oversee artillery posting requirements by discussion with the Indian Army officers and also the Viceroy's commissioned officers. Arrangements have been made for your briefing, after which you will visit, by air, all units in the command, and all the Training Centres. You will be based for administrative purposes on 3/3rd Madras Regiment, Bellary. You are to be gazetted Captain forthwith with the designation Artillery Liaison Officer. Good luck in your new job.'

The size of the job was immense. He was to travel the length and breadth of the Command and to become a guru on the castes and religions of recruits to the Artillery. He would be surrounded by books and other information and appeared to be digging himself deeper and deeper into his life out East. What would he tell Betty?

Betty's letters were now coming through addressed as 'Captain', but he wondered what trick of fate had landed him in this appointment. On the one hand he yearned to get Home to her. On the other, he was captivated by India and its people.

Bill was almost too busy to notice how quickly 1944 was passing. He took and passed his Higher Urdu exam and was receiving additions to his captain's pay both for that and for his Tamil qualifications. Then, quite suddenly, there was an order from army command to stop recruiting. All units were fully complemented.

Then Bill received a telegram from GHQ. His repatriation had been granted. The order was signed 12 June 1944, which by coincidence was exactly five years since he had left England. He just could not believe what he read. He sent Betty a telegram:

'Repatriation granted. Leaving shortly. Adoringly Bill.'

Two days later, he had her reply. 'Absolutely wonderful darling. Longing your return. Adoringly Betty.'

He was 'in the chair' for drinks all round in the Mess that night.

9

LOVE — TREASURE BEYOND PRICE

It was a whole day before Bill came back to earth. He application was stamped and returned. He was to report at the Transit Camp at Bombay on 1 July, to await a passage. He should be Home sometime in August.

He wrote to Mrs Spender, whose address he had in Bombay, saying that he was away. There wasn't enough time to ask Betty what she would like to be brought back. He asked Nora Spender to imagine that she was a young girl, about to be married, and to consider what she would like brought. The only limitation being the size of Bill's trunk! He said that he would call as soon as he got there, and perhaps they could go shopping.

When he met Nora Spender in Bombay, Bill thought that she was about Betty's size, so they went from shop to shop. In the end they had some beautiful dress lengths of Chinese silk, skirt lengths of tweed, blouses, stockings, underclothes, and at Bill's choice, a pair of exquisite wedding shoes hand-embroidered with coloured silk dragons and leaves. Now all he had to do was to get them through the customs. He took Nora out to lunch afterwards and gave her a kiss when he left. He was sure that the Colonel had had a hand in his release.

The *Georgic* sailed on 4 July. She had had a rough time and had to be salvaged after a torpedo attack. She steered somewhat crabwise, none of the doors shut properly, but she was taking Bill *home*. They had to wait at Port Said to be escorted, with other ships, by two frigates.

Bill had to keep a tight grip upon his excitement. Five years since he had seen Betty; with all her experiences, would she have changed? And what about him? After five more years in the tropics he was not now quite the exuberant young man as when they were in Austria together. Cooler now, he was glad of his battledress.

Then the first sight of England. It was dark when they came up the Channel, but Bill was up at daybreak to see the cliffs of Dover peeping through the early morning mist. He felt a lump in his throat; what a bulwark she had been while he had been away, and what about the desperate fighting that was now going on on the second front? It was late afternoon by the time they docked at Tilbury. An army truck whisked the gunners among them away to Woolwich; it didn't occur to Bill until later that there had been no attempt to examine his luggage containing the presents for Betty!

An orderly met them and took them to a reception desk. Bunks for the officers in 'D' wing. Re-registration in 'A' wing 0900 hrs tomorrow, 14 August 1944. Officers are advised to stay in quarters overnight owing to irregularity of public transport. Bill dumped his kit on the bunk and went off to find a phone. Lifting the receiver, 'Bagshot 2386 please,' he said without having to refer to the directory.

A voice answered: 'Hello?' it said.

'It's me, Bill.'

There was no reply – just the sound of the receiver being dropped on the table and a high-pitched scream: 'Betch, Betch, Betch – it's Bill! It's Bill!' It had been Little Ruth, now not so little, but still only nine.

'Oh darling, how wonderful – I just cannot believe it!' Betty's voice was like an electric shock tingling down his spine.

'No, darling, nor can I, but it's me, and I'm here. Oh, it's just wonderful to hear your voice.'

'Where are you?' said Betty.

'At Woolwich. We are being 'booked' out tomorrow morning. I ought to be down by lunchtime.'

'Oh, Bill, can't you come down now? – the bombing at night is terrible, particularly over east London.'

'I wish that I could, Betty, but we have been advised to stay in tonight. Are the Bagshot trains running?'

'Oh yes, I think so.'

'Then I'll ring you from Waterloo tomorrow to say what train I'm catching. Oh Betty, I do love you so.'

'Bill, do look after yourself tonight.'

Bill went back to his bunk and threw himself down on the blanket with his thoughts. The doodlebugs and rockets were just starting.

Re-registration was a peremptory affair. His Army number, his previous unit and location, his address in England. He was given ten days' disembarkation leave and would be told when and where to report to his new unit. Then he was sent along to the MO's office for medical examination. It was just after 11 when he rang Betty again.

'I'm on the 11.25 for Bagshot, darling.'

'Bill, Mummy has made us a packed lunch. I'll meet you at Egham, on the way down. Then we can lose ourselves for the afternoon.'

'Oh, darling, what a wonderful idea.'

The next few hours were forever emblazoned upon Bill's mind. Betty was standing by a little platform shelter, a basket by her side. She wore a summer dress with large pink and green printed flowers on a white background, a white belt around her waist, white leather sandals. Her arms and legs were bare. Her fair hair caught the sun as she threw out her arms to him. As they clasped each other neither could speak – the moment belonged to Heaven. Eventually, holding her forearms: 'Let me look at you, Betty – how many times have I dreamt of this, our first meeting.'

Releasing her, he walked over the sleepers on the line to find a trolley for his trunk. He was conscious of her watching him; it was as hard for her to believe what was happening. When the trunk was safely deposited, Bill turned to her.

'Now, darling?'

'Bill, I'd like to go to Egham church. We have so much to be thankful for.'

Bill took the lunch basket and they walked, hand in hand, a young couple lost to the world in their happiness. They knelt in a pew for a time before walking up to the altar to give thanks to Him that, despite all the odds, they had been brought together once more. It was a solemn and unforgettable moment. 'And now, darling, let's find somewhere quiet where we can just be alone together,' said Betty, taking his arm as they walked down the aisle.

They walked out of the town and slowly climbed up Cooper's Hill. Jumping over a gate, they found a grassy patch in a field overlooking the ancient site of Runneymede where Magna Carta was signed. 'What a lovely spot,' said Bill as he took off his battledress jacket for their heads as they lay on the grass holding hands and gazing up at the sky. It was one of those perfect summer days, warm with the sun shining but with a nice refreshing breeze.

'How quickly it all happened in the end, darling,' said Betty. 'I just couldn't believe your cable from Bombay. We have all been in turmoil ever since!'

'I'm not sure that I can believe it all at this very moment, Betty. Have you got a needle or something to prick me with?'

'No, you idiot,' she laughed. 'There, perhaps that will wake you up!' rolling over to give him a kiss on the mouth.

She undid the picnic basket. Delicately, she laid a tablecloth on the grass.

As she did so, her blue sapphire engagement ring caught the sun. He clasped her hand and gently took the ring off her finger. Then looking straight into her eyes he said: 'Darling I love you. Will you marry me?'

'Oh, Bill. You sentimental old thing. Yes, I adore you and want to be your wife.' Then he quietly returned the ring to her finger, where it had been for the last four and a half years. They kissed each other passionately once more.

Mrs Francis must have used nearly a week's rations on butter, cheese and biscuits. Then there was an apple each and some of her raspberry wine. There was also a flask of tea and a little milk.

'How long a leave have you, darling?'

'Only ten days, I'm afraid.'

'Only ten days after all that time?'

'Yes; we must make the most of it. By the way, the Depot gave me a book for a fortnight this morning. Here it is,' he said, handing over the ration book.

'Mummy has suggested that we should go down to Torquay for a few days; she will come with us as chaperone. She knows a small hotel where we can stay; in any case she needs a holiday.'

'That sound a splendid idea, my precious; when can we go?'

They were both loath to leave their picnic spot – there was so much to talk about. More importantly, they were just 'together'. For how long had Bill yearned to have someone to share his life, his thoughts, his joys and sorrows, and his experiences. Now, by some miracle, his loneliness was to be at an end. He kept his thoughts to himself, but held her hand silently as they rose to go.

All the family were there to greet them on their return. It was much the same as he remembered from his leave in 1939. No, a little more tense and purposeful, but the same family atmosphere that he liked so much. Then Betty took him up to his room and one of the boys carried up his trunk. 'Sit down on the bed,' he said to Betty, 'I have some things for you.'

Mrs Spender's choice was impeccable. Betty gasped as each item emerged from the trunk. Dashing into her room she held up the silks across her shoulder in ecstasy. 'Oh! Bill, I am speechless,' she said. Finally, there were the wedding shoes. So far, somewhat coyly, they had not actually mentioned their wedding to each other; perhaps they were both sensible enough to realise that things must not be rushed. But here was something real in relation to it. Picking up the shoes, and clasping them to her bosom, she ran to her room once more. 'You must not see me in them, must you?' she laughed. She returned – 'They fit perfectly, darling. Bless you.' There was a flushed colour to her cheeks that he had not seen before.

'Where *did* you get all these lovely things?' she asked.

Bill recounted his shopping spree with the Colonel's wife in Bombay.

'She must have such good taste – or was it, by any chance, partly that of my future husband?'

Then there was a fashion parade downstairs, with many 'Oos' and 'Ahhs' on all the items, with the exception of the wedding shoes, which remained firmly in the bottom drawer in Betty's room.

The three of them duly departed by train for Torquay. It was a strange trio: two young people madly in love and the mother 'to see fair play'. Mrs Francis spent much of the time in the hotel, or shopping in the town, while the young ones went swimming from the beach or for long walks together along the cliffs. After three days, Mrs Francis returned home. Evidently she had decided that her future son-in-law could be trusted, or, more likely, Betty had had a word with her. Having been true to each other for so long, they were hardly likely to throw that trust away now. But they were soon making up for years of separation; there was togetherness in everything, time and place just did not matter.

On their last night they went to a ball at the Imperial Hotel. Betty looked absolutely stunning in a long crêpe-de-Chine turquoise evening dress with a rose in her hair. Bill had somehow managed to preserve his dinner jacket and was relieved to find that it still fitted him. It was a perfect evening; as they sat out on the terrace overlooking the bay and watched the setting summer sun, Bill felt an uncontrollable joy within him. It was quite indescribable, as if he was being lifted up to eternity. He put his hand across to Betty and put his head upon her shoulder.

On their return to Bagshot they found that Bill's posting instructions had come. He was to join the 25th Light Anti-Aircraft Regiment in Kent the following week. There was one other thing that he wanted to do during his leave. He and Betty would use some of his 'leave petrol' to see his Uncle Tun and Aunt Hilda at Harwell. They were invited for lunch and tea. It would not be an easy visit. Their only son, Val, who was a pilot in the RAF, had lost his life on a bombing raid over Germany about six months ago. They were alone and both ailing.

Bill was shocked to find them so aged since his visit in 1939. They had both been devastated by Val's death. Their motions and speech seemed almost mechanical, as if their minds were numb. Bill did what she could to give them some feeling of purpose. Betty gave examples of others who had risen above grief, but her efforts were just politely received. Bill was saddened by the visit. After all, Tun was his father's brother, and the only close relation whom he now had in England.

Bill was strangely silent as they drove back. Eventually, she put her hand across to him. 'Is there anything the matter, darling?' she said.

He did not reply at once, then: 'Do you know, Betty, sometimes I feel thoroughly guilty at being alive.'

When his leave ended he joined his unit. The 25th was a battle-scarred regiment that had been through the London Blitz. Their equipment was more modern than anything Bill had experienced before. They were in action on his first night when three Bugs came over. The radar had given warning and two were exploded in the air, giving a ball of fire that lit the sky. The other was 'winged' and came down near one of the gunsites with a sickening thud followed by an explosion.

This continued during August and September. One night Bill was at the Command Post. Some distance ahead they could see a flying bomb coming straight towards them. At least four of the battery guns were engaged but incredibly it survived. The battery behind them opened up, the shells skimming over them. The Bug was now losing height, caught now by the searchlights, its evil-looking features clearly visible. Bill watched the thing as it passed barely fifty feet above them – a long cigar-shaped projectile with single squared wings, the jet engine in its tail spluttering as it ran out of fuel. It flew on for two or three seconds before hitting the ground with a rumbling explosion.

It was Betty who first mentioned a date for their wedding. She had missed Bill terribly while he was away, even though he tried to ring every evening. 'Darling, as soon as possible,' he replied.

'Then what about 21 October?'

'Oh, Betty, how wonderful! I shall have to get leave, of course . . .'

It would have to be a small wedding owing to wartime restrictions: just Betty's family, Avice and Kathleen and Bill's uncle and aunt. Osmond MacDermott would be best man. They saw the Vicar of Bagshot and Bill cabled his mother. So sad that she was still in Canada.

Bill arrived at Park Hill on the evening of 20th to an atmosphere of excitement bordering on chaos to be met by Joan, who had just ordered Betty's flowers. Heavens, he had forgotten all about that! However, he had got the platinum wedding ring, together with a sapphire eternity ring for her wedding present which he had bought in Ipswich. The wedding presents were all out on the dining-room table; they had been eating in the kitchen for the last two days!

Then Bill was solemnly, and rather unwillingly, shooed out of the house. It had been arranged that he would stay the night with some friends: a young couple with nine- and eleven-year-old boys who were, at first, respectfully impressed at having this young officer staying in the house. Osmond had arrived and was staying at The Cricketers, so Bill joined him for dinner. A bachelor, Osmond was full of advice, mostly negative, about marriage. Whether he was genuinely concerned for Bill's future or just plain jealous was another matter!

The rest of the next day became a mystical dream. He remembered Betty coming down the aisle on the arm of her father and he caught a glimpse of Mrs Francis' face as she came; her presence beside him in her long flowing white dress with the net drawn over her face; of the Vicar's voice:

'Wilt thou have this woman to thy wedded wife, to live together after God's ordinance in the holy state of matrimony? Wilt thou love her, comfort her, honour and keep her in sickness and in health . . .'

Of placing the ring upon her finger, of receiving the blessing upon their marriage . . .

There were photographers outside the church and the local Press publicised the romance of Betty's shipwreck and Bill's eventual homecoming. He remembered the thrill of being man and wife alone in the taxi to the hotel, receiving the guests and the speeches afterwards. He heard his own voice thanking Betty's parents for looking after her 'while he was unavoidably detained elsewhere', to be followed by Osmond complimenting Little Ruth as bridesmaid, and making some rude remarks about the bridegroom and some very glowing remarks about the bride. Finally, the Vicar spoke about the path of True Love. Emotionally exhausted, they fell into the train for Dawlish finding a corner in the half-empty carriage. Clasping each other's hands they fell asleep.

Of the journey neither had much recollection except of the warm glow of happiness permeating throughout their bodies. At the station a taxi took them to Lady Browning's cottage, just off the sea front. A caretaker opened the door to them. 'Captain and Mrs Baker?' she said, rather coyly looking at Betty's left hand. 'Do come in, dinner is waiting for you.' Lady Browning had left a pheasant for them together with a bottle of wine. Luxury indeed.

After dinner Bill suggested that they had a turn along the front before going to bed. It was a calm autumn evening, quite warm, with the sea lapping the shore. The moon, just past its first quarter, cast a gentle light. They looked at the stars. Just as they had done before, they followed the points of the Plough, then the Great Bear, '. . . and there's the Little Bear,' they both said together. They were holding the rails on the front; Bill put his hand across searching for the ring finger.

'I don't know whether you believe in telepathy, darling, but I do believe that the darned Bear's ears must be burning!'

Arm in arm they strolled back to the cottage. Mrs Jones, the caretaker, had departed. They had the place to themselves. Betty unpacked their suitcase, placing her nightdress and his pyjamas on the bed. As she turned to him, Bill took the eternity ring from his pocket and gently removed her engagement ring before pressing the eternity and engagement rings back onto her finger.

'Oh Bill, how exquisite; you absolute darling.'

'Well, Betty, that means what it is, my precious.'

Betty disappeared along the passage. When she returned: 'It's all yours, cheri,' she said.

When Bill rejoined her, she was sitting in front of the dressing table, her silk dressing gown thrown loosely over her shoulders. As he approached, she got up and turned to him.

'Oh, Betty, I have had no idea how bea . . .' She smothered his words with her lips.

Next morning, as he awoke, he put out his arm over his wife drowsing beside him. 'Darling,' he said.

'Yes, darling?' was the sleepy reply.

'Oh! nothing, darling; just 'darling', darling!'

POSTSCRIPT

Bill and Betty never returned to India or Ceylon. With Betty's help, Bill qualified as a Land Agent and Chartered Surveyor. They had two children, James and Rosanne – a perfect 'pigeon pair' to complete their happiness.